Charles Richardson

Alcohol

A defence of its temperate use

Charles Richardson

Alcohol
A defence of its temperate use

ISBN/EAN: 9783743373655

Manufactured in Europe, USA, Canada, Australia, Japa

Cover: Foto ©berggeist007 / pixelio.de

Manufactured and distributed by brebook publishing software
(www.brebook.com)

Charles Richardson

Alcohol

ALCOHOL;

A DEFENCE OF ITS TEMPERATE USE.

BY

C. GORDON RICHARDSON.

LECTURER ON CHEMISTRY IN THE ONTARIO VETERINARY
COLLEGE.

WITH AN

APPENDIX.

CONSISTING OF EXTRACTS FROM THE PUBLISHED WORKS OF
EMINENT PHYSICIANS AND OTHERS.

PUBLISHED BY
THE NATIONAL LIBERAL TEMPERANCE UNION
TORONTO, 1888.

ALCOHOL;

A DEFENCE OF ITS TEMPERATE USE.

—BY—

C. GORDON RICHARDSON,

LECTURER ON CHEMISTRY IN THE ONTARIO VETERINARY
COLLEGE.

WITH AN

APPENDIX,

CONSIST.NG OF EXTRACTS FROM THE PUBLISHED WORKS OF
EMINENT PHYSICIANS AND OTHERS.

PUBLISHED BY
THE NATIONAL LIBERAL TEMPERANCE UNION.
TORONTO, 1888.

OBJECTS

Liberal Temperance Union.

✳

1.—The objects of the Union shall be the defence of individual liberty against undue state coercion, the protection of the rights of property against unjust encroachments, the education of public opinion in the principles of just, moderate and reasonable legislation as opposed to arbitrary and oppressive measures, the practice of strict sobriety by the members and its encouragement in others, the discouragement of the use of distilled liquors as beverages, the advocacy and support of a license law discriminating between distilled and fermented liquors, and the securing of an effective system of inspection to prevent the adulteration of all alcoholic drinks.

2.—It shall be the duty of the Union through its officers and members to endeavour to induce sobriety amongst those with whom they come into contact and to reclaim persons addicted to the excessive use of intoxicants, to enlist licensed vendors of intoxicating liquors upon the side of temperance, to prevent by all proper means the issue of licenses to any but respectable persons, and to disseminate by literature and in other ways the true teachings of Scripture and science on the subject of temperance.

3.—The Union recognizes the duty or expediency, in many cases, of total abstinence, and a pledge of total abstinence or any other pledge which does not violate the principles of Christian temperance, shall be amongst the means used in combating intemperance. The declaration of adherence to the objects of the Union shall constitute an obligation to practice sobriety.

PREFACE.

The tendency to sensual indulgence is so strong in many, the evil effects of alcoholic intemperance so deplorable, and the probability is so great that some will make the knowledge that alcoholic drinks are often beneficial an excuse for their own excesses, that many who recognize in alcohol a good and merciful gift of Heaven, may be inclined to question the expediency of disseminating scientific testimony to its value. We cannot but think, however, that a general recognition of the truth in regard to alcohol will result in good, that the removal of errors which more or less obscure the minds of intelligent people will permit of more rational and effective methods of diminishing intemperance than now prevail, and that in this respect, as in others, there can be no essential antagonism between truth and goodness. If the desire for stimulants be a proper and natural instinct under many conditions inseparable from the ordinary life of man, as our greatest physiologists believe it to be, and not a sinful and depraved appetite resulting from self-indulgence, it is foolish to attempt the impossible task of eradicating it. Rather let us recognize the value of the instinct, seek to bring it under the control of reason and moral principle, and to remove, as far as may be, the conditions which tend to make it too powerful. There is reason to think that through a misapprehension of physiological facts, coupled with a moral error of no slight significance, a warfare is being carried on rather against alcohol than against intemperance.

While this misapprehension exists amongst citizens earnestly desirous of reducing the sum of human evil, the danger is great of final failure in the warfare against intemperance, and that what measure of success may be attained in regard to this particular aspect of wrong-doing, will be accompanied by evils which more than counterbalance all the good effected.

There are weighty reasons why the public should hear something more than the excited exaggerations of the evils of alcohol that are current. The other side, the benefit of alcohol, should also be presented. A widely spread moral error has gradually come to pervade society, an error having a most important influence on conceptions as to the true essence of morality and the methods and ends of moral government. The most obvious manifestation of this error is the teetotal system, which appears to regard wine as essentially evil, or at least endorses the extraordinary notion that men should despise and renounce a gift of Heaven intended to conduce to their pleasure and benefit, because some in following out the evil in their own natures choose to abuse it and entail on themselves and others the suffering which is naturally the result of wrong-doing. This doctrine, propagated by enthusiastic adherents and of late, there is reason to think, by many who find its advocacy to be a marketable commodity or a means to notoriety and position, has led to a most undue depreciation in public estimate of the value of alcohol; to the loss of many lives, and, on the part of clergymen and others who might reasonably be expected to practice a little honest investigation, to absurd and harmful exaggerations of the evils of drink. The teetotal doctrine must have indeed become a strong delusion, when prominent men are found to assert that one-third of the earnings of the people is spent in liquor; when symptoms and causes are so confounded that most of the insanity and nearly all the pauperism and crime are referred to strong drink. Because a large number of men, possessing so little moral principle and self-control that they commit assaults and other petty offences against society are also lacking in self-control in regard to the use of liquor, it is frequently concluded that alcoholic intemperance must be, not the symptom, but the cause, of their immoral condition; though a little reflection would lead to the opinion that it would be anomalous if the majority of the criminals referred to did not, as the outcome of their natures, resort to self-indulgence in alcohol.

Still more strange is it to find, despite repeated exposures of the absurdity, the statement reiterated again and again that from seven to ten thousand persons in the Dominion annually "sink into drunkards' graves." As deaths from alcoholic excess are very rare amongst women or persons under the age of twenty-one years, and as the mortuary returns indicate that less than twenty-one thousand men die annually from all causes, it seems remarkable that even thoughtless propagandism can repeat from pulpit and platform a "stock" assertion of the prohibitionists which really means that one man out of every two or three in the Dominion dies from the effects of alcoholic intemperance.

Less easy of detection by the majority of intelligent men is the untruthfulness of the teetotal claim that the opinion of medical authorities is all but unanimous in favor of universal total abstinence. That this assertion should impress the public mind is not surprising in view of the active and widespread agencies constantly at work in disseminating it. A reference to the opinions of acknowledged authorities, especially of those of the past ten or fifteen years, will convince the reader that medical opinion, notwithstanding the influence the moral errors of teetotalism have exercised for more than a generation, does not conflict with the experience and observation of ages and the testimony of Holy Writ that alcohol is a blessing to the human race. Wine, which "cheereth the heart of man," it will be seen is acknowledged to serve the no less important purpose of being a most convenient, and often indispensable, agent in promoting his health amidst the many depressing influences incidental to life. To imagine an agent which could so well fulfil these purposes without producing serious mischief when used in excess may be left to the scientific knowledge of prohibitionists, and to their powers of conceiving a wiser plan than has been adopted in the construction of the universe. And how to secure a proper subordination of the love of pleasure to the sense of duty without attaching to an agent, so capable of useful services in cheering the heart amidst the many cares and worries of life, the tendency to inflict severe penalties on those who, by abusing it, attest themselves "lovers of pleasure rather than lovers of God," may be left to the overwise and somewhat materialistic philosophy of teetotallism.

The doctrines of teetotallism have not only resulted in errors calculated to injure the physical well-being of man, and to retard the progress of temperance : their practical outcome has been the establishment of a galling ecclesiastical and social tyranny, engendering widespread hypocrisy and the deterioriation of politics. A state tyranny has also been established which violates many of the liberties which the English-speaking race has purchased by oft-repeated and liberal sacrifice of patriotic blood. A law is in force in many parts of the Dominion which brands as criminal a class of citizens engaged in an occupation which neither the laws of God nor the general conscience pronounces wrong. It subjects them to confiscation of property, and to deprivation of the means of livelihood for themselves and families, and to fine and imprisonment for the simple offence of selling that which the founder of Christianity did not think it wrong to use. It puts a cruel strain upon the domestic relationship which lies at the foundation of good government, by making a wife a compellable witness against her husband ; it forces upon the defendant the strong temptation to commit perjury in order to secure himself from the penalties of the law, and even when he can honestly establish his innocence, the constitution of the Court is such that his testimony may be rejected as untrustworthy against the evidence of hired informers. It leaves the home open to inquisitorial invasion at the mere instance of a suspicious or revengeful zealot, and has led in a number of instances to the fining of men simply for having liquor in their houses for their personal use. The ordinary laws of evidence are set aside by this tyrannical Act. It provides that it shall not be necessary for any witness to depose as to the description of the liquor sold, or that any consideration was paid therefor, or that it was sold to the witness' certain or personal knowledge ; it leaves it to the option of a magistrate, usually chosen not for his impartiality or legal knowledge, but because of his extreme zeal in the cause of prohibition and his likelihood of convicting suspected offenders, to decide whether the case is sufficiently established to call for the testimony of the accused, whom, in default of his rebuttal of the often scanty evidence brought against him, usually by hired informers, the magistrate is empowered to convict without either trial by jury or the right of appeal on matters of evidence. And, as if this local option law were not sufficiently unjust, the people of counties averse to it are compelled to bear a portion of the cost of enforcing the tyranny upon their fellow citizens. And strange though it may appear, the pulpit often thunders out denunciations of woe on those who fail to support this iniquity.

But the length to which the teetotal tyranny goes does not end here. The teaching of views on the nature and use of alcohol at variance with the general opinion of eminent physiologists—views of a physician so extreme that he holds alcohol to be wholly "powerless for good"—is to be made compulsory in public schools maintained by the taxes of many who utterly reject the doctrines of teetotalism, and unable to receive that extraordinary delusion—the "unfermented wine" theory, believe that teetotal teaching conflicts with the doctrine of the Gospel. Standing, promotion and prizes are actually to depend in part upon the answers given by the pupils being in accordance with the teetotal view, and school boards refusing to submit to such teaching are threatened with forfeiture of their allowance of Government aid. When our public schools are to be made so ostentatiously the means of advancing the political and religious aims of zealots it is surely time that patriotic men, lovers of liberty and justice, and intelligent believers in the Christian faith, should awake to the dangers threatening our liberties.

Where will this tyranny end? Total prohibition of the sale and even private use of alcoholic liquors is the goal short of which it is the avowed purpose of the prohibitionist not to stop. Total prohibition with its attempts to debar the right of every citizen to use liquor in his own home, with its right of search in any house at any hour of the day or night, with its hosts of spies and informers instituting a general inqisition into every man's habits; perhaps, as in the case of an American state, with its interdict on the sale of wine for sacramental purposes, and the restoration, in our boasted age of advancement, of the old days of persecution for conscience' sake!

To plead that these iniquitous interferences with individual liberty are required to suppress evil; that the people who urge and support the tyranny are conscientious and believe they are serving God, is but to urge over again the pleas advanced in past ages in justification of measures of intolerance. Those who were instrumental in the enactment of laws for the imprisonment and death of heretics, were many of them conscientious; the object they believed they were serving was the promotion of true religion and morality ; and it cannot be denied that they could cite numerous instances where liberty of conscience was abused to the detriment of the public morals and good government. And in many instances these laws had the support of a majority of the people.

The misrepresentation of the attitude of science in regard to alcohol has done so much in aid of the growing tyranny that the Literature Committee of the Liberal Temperance Union have deemed it necessary to lay before the public, in convenient form a little of the medical testimony on the other side of the question. The subjoined work, which they commend for careful considera- tion, comprises a valuable essay by an able and careful expositor of chemical science and a painstaking student of the alcohol contro- versy, Mr. C. GORDON RICHARDSON, and necessarily brief selections from the writings of eminent medical men and scientists. The theories advanced by these writers do not in every particular coincide, but there is a significant degree of general agreement as to the fact that alcohol is of much service to the world, and generally conducive to the health and comfort of those who use without abusing it. Sufficient, it is hoped, has also been presented in this volume to warn against improper and excessive use, and to indicate that in some instances total abstinence may be advan- tageous. If the work presented prove to be of value in enlighten- ing some at present misled by the physiological errors of prohibition- ism, in enabling some to realize more clearly that Christianity and Science are not at conflict, in inducing many to take a bold stand for liberty and Christianity against the influences which, under the guise of philanthropy, are tending to subvert both, and in helping to recall public attention to truer methods of moral reform, the compilation of this work will not have been in vain.

THE EDITOR.

TORONTO, January, 1888.

ALCOHOL;

A DEFENCE OF ITS TEMPERATE USE.

—BY—

C. GORDON RICHARDSON,

Lecturer on Chemistry in the Ontario Veterinary College.

CHAPTER I.

STIMULATION AND STIMULI.

To reason clearly, as also to express ourselves intelligibly, it is necessary to acquire the correct signification of the terms we employ. This is most important in the discussion of scientific subjects, where the use of ambiguous phrases, or words of double or doubtful meaning serves not only to retard the progress of scientific thought, but also to the conception and dissemination of false and vicious theories, and to the perpetuation of error.

This latter consequence is amply illustrated in Physiology by the careless use of the word stimuli, where the primary signification of the term has left a strong impression upon medical thought long after the theories upon which the usage of the term was based, have been superseded and abandoned. Not indeed, that the new theories were found more acceptable to the senior generation of medical men, or that the changes were anything but slow and gradual, or that the present generation will be found free from the peculiar bias of former theories. Scientists are not free from that conservatism of thought which shrinks from acceptance of new ideas which are found to clash strongly with the old. It is true that the beclouding of modern ideas in the mists of former theories may in part be attributable to this reason and to the fact that the vast majority of the medical men of to day have received their tuition at the hands of professors impregnated with the color of former theories. Much more, however, will it be found due to the use of words or terms which, while yet applied to phenomena of which our conceptions have materially altered, have not lost any of their original and primary signification.

The word stimuli has an intimate connection with its Greek derivative, στιγμος, a pricking or stabbing, implying also the distinct idea of injury inflicted, and it is in this sense invariably that the older writers used the term. Accordingly we find their definition to have been " a goad,"—anything which like the spur in the side of a lazy horse elicits force while imparting none. Keeping this definition in view we shall see how fitting was its use to the ideas of the older schools, and how fruitful a source of error it has become by careless usage under modern theories through reason of its original significance. The better to understand the Doctrine of Stimulus at the present day, let us engage in a brief retrospect of the theories which have preceded it. It is unnecessary to refer here to the theories of Plato, since they were practically resuscitated in those of Van Helmont and his followers, the impress of whose thought may yet be found lingering in the concepts of modern writers.

Van Helmont regarded the body as the abode of an Archæus, an immaterial spirit (ψυχή) which presided over its functions; that this spirit was subdivisable, the several portions exercising authority over the different members of the body. Perfect equilibrium or *consensus* (Anstie) of these different portions of the Archæus was regarded by Van Helmont as constituting perfect health, while disagreement was ill-health and disease. This idea of a presiding vital spirit was no mere abstract one with Van Helmont, an accomplished chemist, the discoverer of gaseous bodies: he seems to have identified the Archæus with one of these subtle elements. The personal element entered largely into the conception of Van Helmont; accordingly we find the spirit, or spirits, designated under the generic term of Archæus was supposititiously endowed with a capacity for rage, pleasure, or somnolency. It was also considered, although not so clearly, to be a reservoir or generator of energy. We find also the conception of a distinctive action of certain drugs, some being regarded as having an exciting tendency, and others a soothing or subjective action. Van Helmont speaks of "the subjective virtues of carthartic or purgative, the somniferous faculties of hypnotic or dormitive, medicaments." "These words," says Dr. Anstie, "foreshadow the doctrine of *appropriate stimuli.*"

We readily perceive from the foregoing that the Helmontian theory necessitated the assumption that energy was capable of being generated, or created. This was not only recognized and tacitly admitted by Van Helmont and his followers, but was held till very recently to be entirely in accordance with scientific facts by philosophers of high attainments in physical science. It is most important to recognize this fact, *i.e.*, that the older schools admitted or explicitly taught the generation of energy by the vital spirit; because it is upon this assumption that the whole of the doctrine of stimulus was based. Disease then, with the followers of Van Helmont, consisted either of a want of harmony among the various members of the Archæus, an indisposition on the part of one or more of the members to perform their allotted functions, or, in the case of fevers, the introduction of a foreign or toxic body into the system threw the Archæus into an "enragement" who immediately sent into the part affected a peculiar "ferment" which caused irritation of the tissues, calling the blood into them and so inducing inflammation. The *rationale* of the treatment pursued in these different cases then can be easily understood. When, by reason of lethargy in any of the different portions of the Archæus or refusal to perform their allotted functions, that special part of the system became diseased, it was necessary by the introduction of some "spur" or "goad" (appropriate—stimuli?) to arouse the part into proper activity. When the whole of the vital spirit (system) was diseased, recourse was had to the agents known as

" general stimulants." When this introduction of stimulants was carried beyond the point desired, viz., healthy or moderate activity, the Archæus " goaded into fury " gave way to internal heat and passion, causing fevers, tearing and destroying the tissues in its eagerness to dispel the disturbing cause." Hence the signification of the term " overstimulation." The personalty was still more clearly shown in the case of fevers in the use of " somniferous medicaments," and we can, at the present day, easily trace this olden idea in the use of such significant terms, as " allaying the nervous excitement," " nervous irritability."

We have seen that this doctrine of Van Helmont necessitated the assumption that energy was capable of generation. In later times some there were who perceived that this was not altogether in harmony with physical theories, and amongst them was John Brown, the author of what was known as the "Brunonian system," in which he taught that the animal system was endowed at birth with a certain amount of excitability (energy?) the operation of stimuli upon which, while calling forth energy, at the same time exhausts the natural store. It can readily be perceived that while this evades the difficulty of imagining energy as capable of being generated, it in no other way differs from the original conception of the Archæus.

Dr. Anstie has shown that all the leaders of medical thought from the time of Van Helmont till but a few years ago were imbued with Helmontian ideas. There were, of course, many modifications and variations, the most important of which, that of Brown's, we have already indicated, but they all without exception contained the fundamental idea of the Archæus, which, disguise it as they would, still like King Charles' head, crept in.

The first to raise a protest against the soundness of the old theories, was Professor Bennett, of Edinburgh. He showed that so far from the phenomena known as " overstimulation " being a continuance of healthy action to excess, the opposite is really the case, since the results are essentially of a " low " kind. Instead of making the part " more alive " the immediate tendency is to "deaden " or " disintegrate." Professor Lister confirmed these observations of Bennett's, and the remarkable discoveries of Bernard and Brown-Sequard followed in the same direction. These noted physiologists showed that lesions of the vaso-motor nerves are followed in the parts controlled by the affected nerves by phenomena in every way identical with those known as sequences to " overstimulation."

The importance of these facts in assisting the overthrow of the Helmontian theories will be apparent upon a brief examination. According to the old theories energy was a product of the " vital-spirit," distributed to the different muscles of the body by medium of the nerves. If this were true then it follows that any interruption of communication between the source of energy and the vital-spirit, or Archæus [in some of the older writers the vital-spirit is

4

seemingly confused with energy itself] should result in loss of energy. In other words, if motion, energy, animal heat, and all the general phenomena of life are attributable to anything,—vital-spirit, Archæus, or what not—*apart* from the muscles themselves, then any cutting off of communication between them and the source of energy must result in practical "death" of the part affected, loss of power, loss of sensation, incapacity for irritability, and a general sinking of the temperature. Now what are the real facts? Briefly these. On severing the sympathetic nerve of the neck, an immediate elevation of temperature in that portion of the face and neck corresponding to the nerve injured is observed. The skin is hot and flushed. The arteries are flooded and beat with increased vigor. The eye and ear become *over-sensitive* to pain, and the muscles are more easily excited to contraction. All observed facts are antagonistic to the old accepted theories. Again, energy in the animal body was assumed to be a property peculiar to *vital-*force. Yet if this were so, how are we to account for the contractile power displayed by the muscles after death? This [the rigor-mortis] does not set in until all trace of vitality has disappeared, and having once set in does not relax till putrefaction destroys the structure itself. Again in certain nervous disorders, such for instance as paralysis, when all power of voluntary contraction is destroyed, the muscles still possess the power of contraction. These considerations will lead the thoughtful reader to the conclusion that the true action of nerve force appears to be rather in the way of directing and restraining the muscular energy than of exciting or supplying it, and to this conclusion modern theorists are more and more inclining every day.

What then are true stimulants? And how are we to distinguish true stimulation? An answer to these questions we shall endeavour shortly to supply. In the meantime let us recognize with Dr. Anstie that " It is most unphilosophical to speak as if we still believed that a demon or demons resided in our stomach, our nerves, or our ultimate cells whom we could propitiate with cordials, sooth with anodynes, excite with stimulants, or inflame into active wrath with irritants."

CHAPTER II.

ALCOHOL "NOT FOUND IN NATURE."

" The fact that it (alcohol) is not found among the varied compounds existing in animal and vegetable substances in their natural condition, is, of
itself, sufficient to exclude it from the list of necessary articles of food."
—Manual of Hygiene, Chap. XV,, Sec. 49.

The above quotation has been selected, not simply that it
embodies within it a very frequent assertion of teetotal writers and
speakers, but for the reason also, that it forms part of a text book
recently foisted into the Public Schools of Ontario as representing
the latest consensus of scientific opinion upon this question.

But what is meant by "not being found in nature?" Are we
to understand by this purposely ambiguous language, that the
natural laws governing Biology and Botany are not concerned in
the production of alcohol? That it is a mere mechanical production similar to a wagon or a plow? Then the absurdity would surely
be apparent to the simplest intellect. If it is meant that alcohol is
the result of purely chemical action wherein those mysterious
organisms upon the border-land of life, the yeast-cells, have no
part or action other than that which black oxide of manganese
exerts in the production of oxygen from potassium chlorate, then
the statement merely reiterates the theory of Baron Liebig, which
has long since been superseded by the more tenable hypothesis of
Pasteur.

Or are we to understand, according to a more explicit assertion
once made to the writer by a Reverend Doctor of Theology, that
"though God made sugar, the devil made alcohol?" The statement might possess a flavor of wit, but it certainly could not be
truthful. Yet how often assertions similar to the above are made,
either explicitly or impliedly in the vapid utterances of teetotal
writers and lecturers.

But if, ignoring the scientific vagueness of the term "natural
condition," we confine ourselves to the more explicit statement
that it is not found existing in either animal or vegetable bodies,
then I can have no hesitation in declaring the assertion to be
wholly false, and that the author of the above quoted paragraph
has been guilty, either of culpable ignorance of certain well established scientific facts, or else of their deliberately dishonest suppression. This may seem strong language to use, but in this
connection it must not be forgotten that we refer to a work that
presumes to scientific accuracy, issued to the general public by the
authority of a minister of education. To what reason, other than

that of deliberately suppressing all evidence which might tend to weaken the pet theory of the prejudiced writer, can we assign the careful ignoring of all admissions made by even such writers as Dr. B. W. Richardson, or that the works and researches of Pasteur and Chamberlain, M. Lechartier and M. Bellamy, Cagniard de Latour, Schützenberger and Dumas, names familiar in the mouths of scientific men as " household words" have been over-looked and not so much as referred to even. Is it possible that the author was ignorant of the fact that the high reputation of M. Pasteur has been built upon his masterly researches into the nature and action of ferments and fermentation? It may be, for this work (Hygiene) presuming to scientific accuracy completely ignores the researches and conclusions of the greatest living authority upon the subject.

Fermentation, as shewn by these observers, is nothing more or less than a physiological function of cell life in general in which those unicellular organisms such as the genus *Saccharomyces*, have no other advantage over any other living cells in the production of alcohol and carbon dioxide from glucose than of manifesting these reactions in a more marked and intense degree.*

All cells, vegetable or otherwise, possess this property of decomposing or breaking up glucose into alcohol and carbon dioxide, and it would therefore seem that in the ordinary process of life where starch is first converted into sugar and afterwards oxidized to carbon dioxide and water that alcohol must of necessity be one of the intermediary products. In such a case alcohol should be a concomitant of all vegetable life in a greater or less degree, more marked in those plants and fruits where glucose is largely formed. That such is really the case has been incontestibly shewn by M. Pasteur and the observers already quoted. That alcohol should only exist in *small* quantities in fruits, etc., is nothing more than what might be expected from the fact, noted by Schützenberger, that these same unicellular organisms possess the property of burning alcohol to carbon dioxide and water under certain conditions which obtain while the fruits, etc., remain on the parent stem, or in contact with free oxygen.

Lengthy and careful experiments have incontestibly proved this production of alcohol by fruit cells even in such large quantities as one per cent. of the total weight of the fruit, and under circumstances which entirely precluded the action or presence of any adventitious ferment. These cases have been where the fruit has been carefully shielded from free oxygen and apart from the parent stem.

But vegetable cells are not alone in possessing the capacity for producing alcohol from glucose ; animal cells also possess this peculiar property, as also the capacity for reducing the alcohol so formed to water and carbon dioxide.

In reference to the assertion that "alcohol is not found in

*Fermentation. Schützenberger Int. Science Series.

animal bodies in *natural* condition," Dr. Richardson admits that it most decidedly *is* found, and that, not only in the natural condition, whatever that may mean, but in all conditions, and that not only is the human body found to contain it at all times and in all seasons, but worse still, not even the body of a confirmed tee-totaller is found free from the "accursed thing." In his Cantor Lectures, 1874-5, Dr. Richardson says that " In plain words Dr. Dupre's discovery suggests that no man can be in strict scientific sense, a non-alcoholic, inasmuch as, will he n'ill he, he brews in his own economy, ' a wee drap.' It is an innocent brew certainly; but it is brewed, and the most ardent abstainer must excuse it. ' Argal, he that is not guilty of his own death shorteneth not his own life.' The fault, if it be one, rests with nature, who, according to our poor estimates, is no more faultless than the rest of her sex."

From the Doctor's innocent reflections upon this " innocent brew " may we not infer that the teetotaller having maliciously ignored the lessons of reason, Science herself had taken him in hand, and showing his "corporation" to be a natural brewhouse, proclaims him the butt and ridicule of all sensible people ?

But Dr. Richardson is not alone in admitting the truth of our condition, *i.e.*, that alcohol is the result of a perfectly natural process. Dr. Norman Kerr, another " gieat authority," (*vide*, all teetotal writings), states that " fermentation is a natural process," and to teach those of our opponents who habitually play fast-and-loose with words the right meaning of certain terms, adds that "*all* wine, whether fermented or unfermented, is an *artificial* production." We can accept this definition as being in accordance with scientific and ordinary usage, and it has the merit of clearness. The bread we daily eat is an artificial production, obtained by the aid of natural laws, yet how absurd would it be to say that the starch to which it owes its nutritive qualities is an unnatural product. We recognize the fact that starch exists as such in different grains and plants, and it is only by the most lengthy and tedious of processes that we separate and prepare it for table use. When so prepared we call it, and rightly, an artificial product. It has been shewn that alcohol exists as such in vegetable and perhaps in most animal organisms in more or less quantity, that it is an intermediary product : the phenomena of life and a concomitant of certain cellular energy. But while alcohol *per se* is found like starch widely distributed in nature it is only by lengthy processes, very much more simple, however, and uncomplicated than is the manufacture of bread, that it is prepared for our domestic uses in the form of wine, beer, etc.

Lunatics are said to reason logically from false premises. What would be thought of the assertion that because bread is an artificial product it should therefore be flung aside as not necessary to the sustenance of life ? And what must be thought of the theory of those who, for precisely the same reason, would reject the use of wine or beer ?

CHAPTER III.

ALCOHOL AS A FOOD.

No part of the alcohol controversy has excited greater interest or elicited more discussion than has the question of the food value of alcohol. The assertion that alcohol is not a food—a very favorite one with teetotallers—rests upon an old hypothesis of Baron Liebig. The great chemist divided the ingesta into two great classes: the nitrogenous and the non-nitrogenous. The former he termed the "plastic elements of nutrition" and considered them true foods in regard to their being tissue-forming or "histogenetic" material. The latter he termed the "elements of respiration" or as they are now termed "calorifacients," whose only function was, according to Liebig, the production of animal heat. That this classification was unsound was soon shown, as in many cases true histogenetic material can be utilized for the production of heat.

The doctrine that only those principles which contained the element nitrogen could be considered true foods was based upon the assumption, by Liebig, that all muscular action involved the destruction of muscular tissue. For instance when a weight is lifted by the hand a certain amount of the muscular tissue is disintegrated or destroyed to supply the force expended in the effort. Now, since all the bodily tissues other than adipose are, according to Liebig, built up of principles containing nitrogen, any loss or disintegration of tissue must be replaced by principles containing this necessary element, and as alcohol contains no nitrogen Liebig could not consider it as possessing any alimentary value.

This doctrine was generally accepted, and for some few years was held to be in accordance with scientific facts. Such was the high standing of its author that, though not based upon any experimental data, it formed the standard by which the nutritive value of any food was judged. Gradually, however, experimental enquiry has demonstrated its error, and it is now completely rejected. And for this reason. If muscular action is consistent with, or involves the destruction of muscular tissue, then the product of destruction or decomposition of such tissue must of necessity be eliminated from the body, and the amount of such elimination should bear a direct proportion to the amount of energy expended. The principal channel by which decomposed nitrogenous matter is eliminated from the system is the kidneys, through which it escapes in the form of urea. Now although, as we have already stated, there were no experimental or analytic data, Liebig and his followers were wont to assert that the amount of urea eliminated increased in direct proportion to the amount of energy expended. It is unnecessary here to detail at length the numerous experiments conducted at various times in different countries and by the most eminent scientists, to decide this important question, since they may be readily found in all physiological works, especially those of Fick and Wislicenus, Parkes and Wollowicz. Suffice it to say, that the experiments all went to prove in the most decided and conclusive manner that so far from increasing in proportion to the energy expended, the amount of urea was in many cases diminished. They amply proved that muscular-tissue disintegration could not possibly be the source of muscular power.

We must therefore look elsewhere than to the nitrogenous material as the source of muscular energy. But Liebig had divided the ingesta into two classes as we have seen, the nitrogenous and the non-nitrogenous. If the former cannot be regarded as the true source of energy then it must of necessity reside in the latter. But how ? The doctrine of the " correlation of the physical forces" enunciated by Grove some few months after the publication by Liebig of his "animal chemistry," had prepared the minds of physiologists for the question. This doctrine maintains that all forces are capable of reciprocal production, or in more simple words, heat, light, sound, motion, electricity and magnetism are interchangeable, that either may produce one or all of them, and that none could originate except from some preceding force or forces. Associated with the " correlation of the physical forces " was the doctrine of the " conservation of energy " which implies that the quantity of energy is as indestructible as matter, and that, however variously it may be transformed in all the manifold changes of the universe, it cannot be created or annihilated, decreased or increased. This doctrine clears the way for a thorough understanding of what constitutes food. By vegetable processes the sun's energy is rendered latent in the carbon compounds formed, and this potential energy is re-converted into actual energy when they undergo oxidation or combustion in the animal economy. This should be apparent to all in these days of electric lighting. Coal is of vegetable origin nourished in the rays of light and heat which left the sun many ages ago. We see it today shovelled into the boiler of the engine and by its oxidation yield up the energy stored up so long amongst its molecules ; the heat is then converted by means of the engine into rotatory motion, this latter is then conveyed by means of shafting, etc., to the dynamo where this common motion is converted into magnetism and electricity, and this again conveyed by means of wires to the lamps which illuminate many of our public streets to be there converted into light and heat and sometimes sound.

In what way then does this modify our conception regarding the nature of true foods ? Simply that we must consider as true foods all principles which undergo metamorphosis in the system, building up the body, or by oxidation yielding the necessary energy for muscular and vital phenomena. Alcohol in Liebig's time was understood to undergo oxidation in the system to carbon-dioxide, dioxide and water. but as we have seen, not containing any nitrogen was not classed as a nutriment. "Alcohol," says Liebig, "stands second only to the fats as a respiratory material." The presence of nitrogen being no longer a test of its food value, alcohol at once assumes a high rank among those foods capable of evolving energy by combustion, and this, not only by reason of its lending itself readily to oxidation but for its easy assimilability, not taxing in any degree the digestive powers.

A reaction from this view, which was generally accepted by chemists upon the overthrow of Liebig's theory, was started by the announcement by three French chemists that they had found alcohol in an unchanged state in the secretions of the body. This question of elimination of alcohol will be treated at length in a succeeding chapter, for the importance of this fact, if true, can readily be perceived. If alcohol escapes from the body in an unchanged state and in any quantity (the words of the French observers were "*en totallite et en nature*"), then according to even recent definitions alcohol could not be classed as a food capable of yielding energy. Suffice it to state now, that these views of the French chemists have been proved to be entirely erroneous, and that the fact that alcohol does really undergo consumption in the body is sufficiently well established as a scientific postulate.

There are some writers with more ingenuity than erudition or even perhaps honesty, who, while yet admitting that alcohol is not in ordinary doses eliminated unchanged, (such as Dr. B. W. Richardson) try to show that alcohol cannot even be regarded as yielding energy since "the temperature of the body is lowered rather than elevated after the ingestion of a dose." Such an assertion betrays a lamentable ignorance of the meaning of the doctrine of the correlation of the physical forces and of even the simplest chemical facts. It is by no means necessary that latent energy should first be converted into heat that any of the other forms of force may be produced, and even were this the case, although it is absurdly false to anyone even slightly acquainted with physics, it by no means follows that the body should be raised in temperature by the large evolution of heat yielded by alcohol during oxidation. That quantities of alcohol capable of yielding very great amounts may be oxidised in the system without raising the temperature one degree above the normal may sound paradoxical to those unaccustomed to questions in molecular physics, but it can be readily seen upon a moment's consideration that such is not only possible but very probable.

In an ordinary locomotive, in a few hours, is burnt a quantity of coal yielding an amount of heat, which, if used for heating purposes would be sufficient to completely fuse the metal of which the machine is made; but the fact is that the engine, as a whole, never rises in temperature much, if at all, above that necessary for the production af the working pressure. What then becomes of the vast quantities of heat generated by the combustion of all this coal? It is converted, as fast as generated, into the mechanical motion which hurries the train to its destination. For every pound carried over every yard of space a certain definite fixed quantity of heat disappears, and for every degree of temperature manifested in the metal of the locomotive just so much energy is *diverted* from application to direct motion. The engine then serves but to direct the energy evolved by oxidation into the desired channels. Just so with the body: in the animal engine the muscles, nerves, limbs, etc., serve to direct the energy evolved during the combustion of the carbonaceous food into muscular and nervous force, only suffi-cient heat being diverted from the various ends as is neces-sary to keep up the animal heat, and no more. But it by no means follows, as before stated, that heat is an intermediate form between latent and actual energy : unlike the crude mechanical devices of man, that intricate apparatus we call the body, beyond doubt converts the latent directly into the form of the actual force desired, thereby saving much loss by friction in clumsy interme-diary appliances.

Admitting all these facts, there are yet some who say : " if alcohol is then capable of yielding energy by oxidation and the nitrogenous material is only necessary for the repair of tissue degeneration, then people ought to be able to wholly live, or at least very nearly so, upon an almost exclusively alcohol diet." Just so. It would not be difficult, since it has already been done, to furnish overwhelming proof that persons not only can, but have lived for days, and months, even years, upon an almost exclusively alcoholic diet ; of course these last are exceptions. What doctor is there, who, if his practice has been at all extensive, cannot recall cases where patients have existed for long periods upon brandy with a supply of other food inadequate in itself to the needs of the body ? All persons could not live in such a way ; much less carry out all the proper functions of vital existence. No single alimentary radical known is capable in itself of properly sup-porting the evolution embraced in the proper carrying out of the ends of living, and just as the ends aimed at are complicated and heterogenous, so will the food supplied to the body need to be varied and heterogenous. It must not be forgotten also that the preparation of food for its assimilation by the body diverts a very large quantity of energy. If then we would preserve as much energy as a certain amount of food ingested will evolve, in order to apply it to the production of higher forms of force, we must (having, of course, due regard to other physiological relations of digestion), cut down the amount diverted in the preparation or *digestion* of such. We can do this by taking *pre-digested* food. Here comes in the valuable nature of alcohol. It is a *pre-digested food*. Of course we are at once met with the suggestion that we should use milk, (*vide* B. W. Richardson's text book on "Temperance.") In the first plac e answer that milk is not a single alimentary radical, but a con. ination of principles, as may be readily proved by allowing it to stand, when some of the ingredients immediately separate out ; and in the second place, although milk, unlike any other food we are acquainted with contains all elements necessary to nutrition it is by no means a *pre-digested* one, as those subject to certain forms of indigestion can testify. Again, let a number of persons for even a short period attempt to perform their ordinary avocations upon milk solely, and many of them will discover that even this " ideal food " is totally inadequate to supply the needs of the average adult, while if we are to believe Sir Henry Thompson, who has much more claim to the term scientist than has Dr. B. W. Richardson, who is ridiculed or ignored by great physiologists, we must accept his assertion that milk is for adults a superfluous and mostly mischievous article of diet: So far as the writer is concerned he could bear testimony to the fact that for himself at any rate milk taken in draughts during or at meals is the cause of a most distressing form of dyspepsia, but there are some to whom a glass of milk at the meals fills a want which no other substance can so well supply. Just as there are many to whom a glass of sherry or madeira fills a need which no drugs in the Pharmacopœa could fulfil, or a still larger number to whom a glass of ale with their dinner constitutes the difference between perfect digestion together with pleasurable performance of their necessary labours, and imperfect digestion with impaired powers of nutrition resulting in languor, dullness, and inaptitude for the cares and difficulties of business.

To all such, and they are very numerous, alcoholic beverages —preferably in the form of well fermented natural wine—are a necessity or at least an advantageous article of diet, supplying as they do in small bulk a palatable, predigested, easily assimilable food.

CHAPTER IV.

ELIMINATION OF ALCOHOL.

" The vital organism obviously treats alcohol as an intruder and irritated by
its presence, is roused to an abnormal state of activity until the last atoms
of the offending article are cast out of the temple which it pollutes. The
body then resumes its ordinary functions, subject to the reaction of
wasted energy."—A teetotal journal.

We stated in our last chapter that a reaction from the general
concensus of scientific opinion, that alcohol was a food as much as
any of the starches or fats, was caused by the announcement by
M. M. Lallemand, Perrin and Duroy, that they had discovered the
elimination of alcohol from the body in an unchanged condition.
Had these observers been men jealous for the honor of science they
would have refrained from making any quantitative deduction from
experiments purely qualitative in their character. They did not
hesitate, however, to declare that they had found that alcohol is
eliminated from the body "*en totalite et en nature.*" The statement
was eagerly received by prejudiced persons and given great pub-
licity for certain easily conceived reasons, but it was immediately
objected to by scientists on the grounds that it was defective in
many respects; first, the want of precision caused by the unknown
delicacy of the test used ; second, that the doses given were always
intoxicating ; third, that it was manifestly absurd to base quantitative reasoning upon qualitative experiments.

The labour involved in a settlement of the question by direct
experiment was great, but soon after M. Edmund Baudot*
placed on record a series of carefully conducted experiments which
in themselves are sufficient to completely overthrow the assertions
of the earlier observers. M. Baudot first notes that if M. M. Lallemand, Perrin and Duroy were correct, a sensible result should be
obtained on an examination of the secretions by means of the hy-
drometer. He then proves that the results in every case under
examination were completely negative. He then examined the
delicacy of the chromic acid test, the test made use of by the
French chemists, found it capable of revealing the presence of so
extremely minute a quantity as 0.155 of a grain to nearly a quart
of urine. We may here note that the principal channel of elimination of alcohol, according to these French chemists, was the kidneys. Now as Dr. Anstie has shewn that the probable maximum
quantity of urine passed during the twenty-four hours immediately

* [Union Medicale, Sep. et Nov. 1863.]

following an even strongly diuretic dose of alcohol is not much more than two quarts, the full absurdity of asserting that because a reaction had been obtained by means of a reagent capable of being affected by about one-third of a grain of alcohol in two quarts of urine, therefore, the whole of the alcohol, even though it were many ounces, as in every instance it was, was totally eliminated, becomes fully apparent even to the unscientific reader.

But the matter was not suffered to rest solely upon the evidence adduced by M. Baudot, soundly scientific though it was. Dr. Anstie whose whole life it may be said has been devoted to the study of "Stimulant Narcotics" directed attention to a lengthy series of experiments conducted by himself which amply sustains those of M. Baudot. He also shews the extreme delicacy of the chromic acid test. Dr. Parkes, by no means an observer prejudiced in the cause of alcohol, in conjunction with Count Wollowicz points out that they have obtained the chromic acid reaction with the condensed perspiration from the arm of a man who had taken no alcohol for twenty-six days previously. Dr. Dupré agrees with Anstie and Thudichum in England, and Schulinus and Baudot on the continent, in believing that the chief quantity of alcohol ingesed is destroyed in the system. In agreement with Parkes and Wollowicz, Dr. Dupré found that after six weeks abstinence from alcohol, and even in the case of a teetotaller, a substance was eliminated in the urine, which gave all the reactions used in the detection of alcohol. "It passes over with the first portion of the distillate," says Dr. Dupré. "It yields acetic acid on oxidation, gives the emerald-green reaction with chromic acid and yields iodoform : " he further adds that this had been previously noticed by M. Lieben. Shortly after the publication of these experiments of Dupré's, an article appeared in the Practitioner (July, 1874) from the pen of Dr. Anstie detailing a number of experiments conducted with a view of settling the question raised as to the possible accumulation of alcohol in the body. In an experiment mentioned therein, brandy was administered to a dog to the extent of one ounce daily for ten days, after which and almost immediately following the ingestion of the last dose of alcohol, the dog was destroyed, and the alcohol contained in the whole of the stomach and body was collected and estimated. The total quantity of alcohol eliminated from the body during the ten days, together with that recovered from the body after death, amounted to only one fourth the quantity administered in the dose of brandy given to the dog just previous to its death. " These experiments," says Dr. Anstie, " furnish us with a *final* and conclusive demon   ion of the correctness of Dr. Dupré's arguments against the   bility of material *accumulation* of alcohol in the body." To tl   justness of these conclusions the most recent writers upon " Foo   and "Dietetics " such as Dr. Pavy and Dr. Milner Fothergill, bear ample testimony; nay more, even the high priest and chief apostle of teetotalism and universal milk, Dr. B. W. Richardson, is obliged, from sheer force of facts, to yield a reluctant acquiescence in the verdict of science, which has completely overthrown the false assertion of the French chemists. In his Cantor Lectures he admits the fact that alcohol is decomposed in the body; and not only this, but that it may be, and is undoubtedly manufactured in the body. Is it not deliciously ludicrous to note the innocent pomposity of Sir Oracle who presumes to quarrel with nature, " who " he says, " according to our poor estimates is no more faultless than the rest of her sex."

CHAPTER V.

ALCOHOL "A POISON."

A favorite assertion on teetotal platforms is the statement that alcohol is a deadly poison. Never will the writer forget the unctuous satisfaction with which a certain lady lecturer enunciated in his hearing that "alcohol is an acrid-narcotic poison," or the sigh of admiring horror with which the statement was received. Were such persons content with the simple statement as given above, no particular harm would be done, nor would scientific truth be greatly wrenched from a straight line. But we must take exception, and that strong, to the logic which asserts that "because alcohol is poisonous in excessive quantities, it must of necessity be injurious in small doses." It would fare ill with humanity were this logic sound, since it can readily be shewn that there is nothing in the nature of an alimentary principle which is not injurious when taken in excessive quantities. We have only to apply the reasoning to other things in everyday use to realize the absurdity of the contention. Salt, an article indispensable to the sustenance of life, is, when taken in excess, a virulent poison. Orfila mentions several cases of death by its agency. Vinegar, mustard, pepper, tea, coffee, all contain principles which, taken in excess, are poisonous, and if the above logic were sound their use would, even in moderation, be highly reprehensible. There are a few who recognize this fact and are honest and consistent enough to follow the logic to its natural end, and who not only abstain from the greater number of the articles mentioned themselves but endeavor to get others to abstain also. Some physicians there are also, who lend a support to their absurd fancies. Not only have vegetarian societies been formed but an anti-salt association, while some few years ago the endeavor was made in New York to found a medical hospital whose prime tenet was abstention from salt, and a work was published by a physician of some local standing entitled "Salt: the Forbidden Fruit," in which it was endeavored to be shown that all the ills to which flesh is heir to were to be attributed to the use of this "most pernicious drug."

To meet this by the objection that "these things do not cause a man to beat his wife, illtreat his children, and generally to ruin his home" is to put forward a statement which is not only silly, but false. The connection between a heavy meal of fat pork, green tea, indigestion, dyspepsia and ill-temper may not be so apparent to those unacquainted with well known physiological facts as the connection between drunkenness and brutality, but the connection is there nevertheless. Even Sir Henry Thompson, with all his teetotal leanings, recognises this fact, and plainly warns us against injudicious eating and excess in all kinds of food, and states that errors in overeating produce more physical evil than excess in drinking, and that he is not sure but the same remark will apply to moral evil also. Homes can be broken up by other means than excess in alcoholic drinks. Evil passions are aroused as easily by errors in over eating as by over drinking, and were more moderation practised in regard to the former, we would doubtless suffer much less from another evil, which drink reformers are too apt to overlook, and which teetotalism is not designed to ameliorate or cure, at least to judge from the experience of Mohammedan countries.

To base any reason for abstinence from alcoholics upon the fact of their being termed " stimulant-narcotics " is equally absurd. Who has not noticed the stimulating effect of a moderate meal taken when hungry, and the disposition to sleep experienced upon taking an inordinately heavy one, and if the tendency to sleep be indulged in, the possibility of rising with a bilious headache. These latter are the narcotic effects of an excessive quantity of ordinary food. So that we learn that the difference between stimulation and narcotism is one of quantity only. " All foods " says Dr. Anstie " are ' stimulant-narcotics.' " This double action, so to speak, is generally recognised, not only in the case of our common foods, bu also in those special foods which are more often admin. with medicinal intent. Again, common food is often administered as a medicine. No more grateful, nor, when it can be properly assimilated, more powerful stimulant can be administered in certain cases than a good bowl of warm soup. So we see that the term " poison " and " food " are simply relative ones, referring to quantities only, and are applicable to all substances used by man in the support or sustenance of vital functions.

CHAPTER VI.

ACTION ON DIGESTION.

Alcohol, or alcoholic beverages, properly used, aid digestion. This fact, attested by the accumulated experiences of past generations, and the daily experience of countless thousands of temperate drinkers of the present, and universally admitted by physiologists so as to have become a postulate almost, in the science of dietetics, has been questioned by teetotallers. It has been said, by these latter, that the corrugating effect of strong spirits, placed upon the back of the hand for instance, has its counterpart upon the coats of the stomach when alcoholics are ingested into the body. To say that the blistering effects of mustard, similarly applied, has its exact counterpart upon the mucous membrane of the stomach, when mustard is swallowed with our food, would be the height of absurdity, yet this is a parallel case. "But," say our objectors, "that cannot be : we only take mustard in such quantities that while helping digestion it can be digested itself." Exactly: temperate people take alcohol in such quantities that while aiding digestion the system can assimilate it. Temperate people are not in the habit of taking inordinate quantities of alcoholic drinks, nor of taking undiluted spirits in the form of drams, nor of even taking alcohol *per se*, a substance, let it be noted, not generally found outside of manufactories or chemists' laboratories.

The local tendency of undiluted alcohol on the living tissues is to impair and finally destroy their vitality by the abstraction of water. They become blanched and corrugated, but with alcoholic liquids taken well diluted, in forms containing not more than 19 per cent. of alcohol, we find the flow of saliva increased. The capillaries of the stomach dilate, its glands at once secrete copiously, its movements are more vigorous, " bringing about," says Lauder Brunton, "a more thorough and rapid admixture of the contents of the stomach with the digestive juices. and facilitating the expulsion of gases." This is beneficial, but when alcohol is taken in an undiluted form or in too great quantity—such a quantity as would narcotize instead of stimulate—the whole aspect is altered, and the secretions are checked. The mucous membrane becomes pale and corrugated, while slimy mucus replaces the healthy gastric juice ; appetite is lost, and nausea and vomiting possibly supervene.

But what lesson do these facts teach us? Not that alcohol properly taken acts injuriously upon the system. Far from it ; but that there is a quantity and strength in these liquors beyond which it would be injurious to pass, just as in the case of tea, coffee, mustard or vinegar. Indeed a much stronger case can be made out against the use of tea as an ordinary beverage, than can be made against the ordinary alcoholic table beverages. It is claimed, in this connection, that alcohol precipitates pepsine. So it does, when in a strong and undiluted form, but not when the quantity of spirit falls below 10 or 12 per cent. Even were that so, the pepsine would not be destroyed, for the moment the contents of the stomach were diluted the pepsine would at once be re-dissolved with all its digestive powers unimpaired. But it is different when we come to deal with the effect of tea upon digestion. One of the most insoluble of compounds known to the chemist is that formed by the union of tannic acid and albumen - known commonly as "leather." Of all substances it is the least digestible. Tea contains a large quantity of this powerful acid; poorer qualities, indeed, possessing scarcely any other active principle than this. If the tea taken into the stomach be properly diluted and of good quality, little harm will result beyond retardation of digestion, which in persons of strong digestion may be often advantageous. But when strong tea is used the effect on digestion is very injurious. Its tannic acid, being then in strong solution. will combine with the albuminoid constituents of the food that are present in the stomach and precipitate them in a very indigestible form. It will even combine chemically with the albuminous elements of the mucous membrane. thus hindering the secretion of gastric juice and bringing digestion almost to a standstill. And even when the mucous membrane has resumed its functions the presence of the now indigestible combination of tannic acid 'with the albummoids will act· like a foreign body and worry and irritate the mucous membrane and cause a feeling of pain, or at least uneasiness, with eructations and marked flatulence, thus disorganizing its functions and giving rise, says a well known physiologist, in Chambers' Encyclopœdia, "to a distressing form of dyspepsia, which too often impels the sufferer to take refuge in ardent spirits." A lady lecturer lately enlightened our darkened intellects with picturesque delineations of a human stomach alcoholized ; might we not offset this by an exhibition of that same useful article in the human economy, tanned by the excessive use of that, which, with ignorant lack of discrimination as to quality and quantity, is often called " good, wholesome tea."

In this connection it would be well to refer to an experiment rather popular with ignorant audiences, and although exposed over and over again, will no doubt be repeated whenever and wherever church pulpits on a Sunday evening are offered for the desecration. Brandy—the strongest—is added to an egg, and when the albumen is coagulated and precipitated it is held up as an example of the action of all alcoholic beverages. But temperate people are not in the habit of taking liquor in the form of strong drams. Again, the lecturer is either too ignorant or else wilfully dishonest, or at the same time he would inform his audience that the cup of strong tea with which, no doubt, some of his lady hearers had just regaled themselves, would have precisely the same effect, or what is more important still, have pointed out the well known physiological fact that the normal gastric juice of the stomach effects the same change : *it coagulates the albumen prior to digesting it.*

There may be individual cases, however, where the doctrine that alcohol favors digestion will seem to be negatived, but a careful examination of such cases will generally disclose the reason of the apparent inconsistency. There are some, chiefly amongst the sedentary classes, who find all forms of beer, even perfectly manufactured beer, totally incompatible with their systems. This is not to be attributed, by any means, to the effect of the alcoholic ingredient, but often to the effect of the hops on the liver. In these cases it will sometimes be found that good spirits, properly diluted with four or five times the quantity of water, and taken after, not before meals, will not only be well tolerated by the system, but materially aid the digestive powers. It will generally, however, be best to use instead of the spirit an equivalent amount of a good wine. There are, it is true, a certain number of persons in this category who complain that wine lies "cold" upon their stomachs. These few, it is to be feared, have inured their palates to the flavor of strong and ardent spirits or else they have been unfortunate in their choice of a wine. A claret or full-bodied, generous Burgundy ought to replace in almost all, if not in every case, the dietetic allowance of spirits. The writer is acquainted with several cases where a glass or two of Chablis has replaced, with marked advantageous results, an equivalent quantity of spirits, and this was the more to be noticed inasmuch as in some of the cases the persons were aged and in the habit of taking spirits for years previously.

For the literary man, delicate women or anæmic children with weak digestive powers and capricious appetites there is nothing so exceedingly well adapted for general use as are the ordinary light, dry wines of Bordeaux. We do not here refer to the thin, sour wines procurable in France under the generic term of *vin ordinaire*, (though even these are to most people wholesome and beneficial), but to the dry-flavored natural wines of the Rhine and Bordeaux. Were our own light native wines, many of which compare favorably with the clarets of Europe, to replace the vile stuff which is vended, especially in the country districts, under the name of tea and coffee, they would be found to conduce greatly to the general health of the community. We do not refer in this connection to those sugared wines generally sold in taverns under the term native wines (doctored it may be with added spirit), but to the naturally fermented juice of the grape without addition of extraneous sugar or spirit. It may be objected that the wines referred to are not procurable in Canada. This may have been the case but it is so no longer. The writer has had no difficulty in procuring a very good native wine of exceptional purity and fair flavor, and at a price within the reach of all who can buy tea. These wines, diluted with an equal quantity of water, form a most grateful and wholesome beverage either for summer or winter use.

A giving up of the use of pastry at one or two of the daily meals, the relegation of tea to the *tea*-table, and the substitution, in many instances, of light well fermented claret in its stead at the principal meal, would, we are assured, do more to banish the doctor for the majority of American and Canadian households than all else besides. Much less, at any rate, would be heard of those intricate nervous disorders of women and children on this continent which test the patience and skill of our best physicians. Much less, also, of that chronic American complaint, but little known in the wine districts of Europe, dyspepsia.

CHAPTER VII

ACTION OF ALCOHOL ON THE HEART.

" The heart of an adult man makes, as we have seen, 73 57 strokes per minute. This number multiplied by sixty for the hour, and again by twenty-four hours for the entire day, would give nearly 106,000 as the number of strokes per day. There is, however, a reduction of strokes produced by assuming a recumbent position and by sleep, so that for simplicity's sake we may take off the 6,000 strokes, and speaking generally may put the average at 100,000 in the entire day. With each of these strokes the two ventricles of the heart, as they contract, lift up into their respective vessels three ounces of blood each, that is to say, six ounces with the combined stroke, or 600,000 in the twenty-four hours. The equivalent of work rendered by this simple calculation would be 116 foot-tons ; and if we estimate the increase of work induced by alcohol we shall find that four ounces of spirit increases it one-eighth part ; six ounces, one-sixth part, and eight ounces, one-fourth part."—*Dr. B. W Richardson, in Cantor Lectures.*

One is puzzled upon reading the above, whether to laugh at its pompous explicitness, or sigh for the ignorance of common physiological facts displayed in the inferences which the author immediately proceeds to draw. Divested of all appendages of pseudo-scientific jargon the reasoning amounts to this : Alcohol accelerates the action of the heart. The heart's action equals so much expenditure of force. Ergo, any increase of the heart's action beyond normal proves the injurious action of alcohol !

Do you indulge in football? You are unwise. Are you not aware that football, even when indulged in to only a moderate extent, greatly accelerates the heart-beats? And if entered into with anything like vim, will send your heart thumping against your ribs as though it would force an outlet through them ? Please not to speak to me of that feeling of general "well-being." That is but the result of the increased blood pressure on the base of the brain. Tell me not of the exhilarating effects of the quickened circulation. Think of the consequences implied in the increased oxidation, the enormously increased metamorphism, and destruction of tissue, and the wear-and-tear of your unfortunate heart. Think of the great expenditure of force or energy implied in those many beats beyond the normal. Calculate the foot-pounds expended in this way during one minute, multiply the minute by the hour, the hour by the day, this by the week, then struck aghast by the hideous length of figures, foreswear football forever.

Football accelerates the action of the heart. The heart's action equals so much expenditure of force. Ergo, secundum Benjamin W. Richardson—increase of heart's action proves the injurious effects of football.

The ardent disciples who absorb the heart-action argument of the worthy doctor should regard walking as a perilous exercise, seriously detrimental to longevity, and should indulge in it no more than the necessity of earning a bare livelihood absolutely demands, and when not obliged to walk or stand should lie down rather than sit, for the nearer the approach to a perpendicular position the more ominously do the extra heart-beats give warning that the heart is being overtaxed, and is wearing out, and the vital energies are becoming exhausted.

The absurdity of such reasoning must surely be apparent to the intelligent reader. Yet this is embodied as science in a text book on alcohol published for use in the public schools. But there are other serious defects in the paragraph which heads this section of our work than the deductions based upon the mere increase of the heart's action. One is the quantity of alcohol used. Now no one can for a moment countenance drunkenness in any form or degree. It would be absurd to attempt to defend the intemperate use of liquor from a scientific standpoint. The physiological evils of intemperance are admitted by all. In the question whether we shall rigidly abstain from alcohol in any form or not, we must argue upon the grounds of moral expediency in each case, the moral right to temperately use, and the nature of the physiological effects of moderate indulgence. In any work purporting to deal with the scientific aspect of the question any consideration of the two former grounds would be apart altogether from the subject. We thus find our ground limited to a discussion of the physiological effects of moderation. This has been foreseen by the teetotallers for many years, and in consequence they have turned their guns upon the temperate drinker. It must surely be apparent to the unbiased reader *that to do so logically and honestly* they should confine their attention *strictly* to the effects of moderation. To say that because alcohol is injurious when used in excessive and intoxicating quantities, it is therefore hurtful when taken in moderate quantities, is to assert that which is both absurd and untruthful. To wilfully confound the effects of *intemperance* with those of *temperance* is wicked. Yet this is continually being done by the advocates of teetotalism, and not the least sinner in this respect is Dr. B. W. Richardson.

In the paragraph immediately following the one we have quoted at length, Dr. Richardson proceeds to tell us " what we may call a moderate amount of alcohol, say two ounces by volume, in the form of wine, or beer, or spirits." Now, what the doctor is pleased here to term a " moderate amount " is that which modern science has shown to be the average *limit beyond which it would be generally best not to pass.* This represents a pint and a half of claret, or about two pints of beer per diem. In passing, I might point out an error which occurs amongst others in this same paragraph : were it not that Dr. Richardson is referred to *ad nauseum* as a scientist by those whose geese are usually swans, the error might be overlooked. The Doctor says that strong ale contains *fourteen* per cent. of alcohol. The fact is that Scotch ale contains but eight per cent., and this is the strongest known in America, while ordinary strong stock ale rarely contains more than seven ; yet the Doctor has little hesitation in doubling this figure.

Now, for the sake of argument, let us admit the evil effect of greatly increased heart action, and keeping in view the fact that we are only fighting the fallacy that presumes to assert that moderate drinking is generally injurious, we demand the effect of moderate quantities. This does not suit, however, the fancy of our worthy scientist, who ingeniously proceeds to double, then treble, and finally, in mathematical agony to quadruple a quantity which, to commence with, we know borders upon a narcotic and intoxicating dose. It is true that immediately following upon the Doctor's definition of a " moderate amount " there occurs this important admission : "When we reach the two ounces, a distinct physiological effect follows, leading on to the first stage of excitement with which we are now conversant. The reception of the spirit arrested at this point, there need be no important mischief done to the organism"—but this serves only to show more clearly the dishonesty of attempting to confound the effects of intoxicating doses with the effects of moderate indulgence. In this short sentence the Doctor innocently gives his whole case away.

"The perils which environ
The man who meddles with cold iron "

are as nothing compared with the pitfalls prepared by science for those who use her language but lack her acquaintance. Mark the significance of the admissions, for there are several, contained in the paragraph. " When we reach the two ounces," in other words, when we reach the average safe limit " a distinct physiological effect follows." We here see that this effect did not exist, according to the Doctor's own words, *prior* to reaching the safe limit, " leading to the first stage of excitement with which we are now conversant." Astonishing, all this only leading up to the "*first stage of excitement.*" " The reception of spirit arrested at this point, there need be no important mischief done to the organism !" This looks, from a teetotal point of view, very much like what Lord Randolph Churchill has expressively called " chucking up the sponge," does it not ? It is rather curious also that the author of the chapter on alcohol in the " Manual of Hygiene " while extensively quoting from Dr. Richardson, should overlook such remarkable passages as these.

But what does science really teach in reference to the action of alcohol on the heart ? A reference to a few of her recognized mouthpieces will I think enlighten the impartial student that the opinion of science may be summed up as follows :—That taken in moderation there is no more action upon the heart than what might and does follow the ingestion of ordinary food, or moderate indulgence in healthy exercise : that the average safe limit is two ounces, beyond which narcotism may set in, and increase with the increasing quantities of alcohol taken : that until such limit is reached there is not generally any injurious effect whatever. Why then, it may be asked, if scientific men are of such an opinion, have I dealt largely in this chapter with the opinions of Dr. Richardson? One reason may be that Dr. Richardson, being the acknowledged head of the medical teetotallers, it is well to show that he is not even personally consistent in his statements and deductions. Again, it may be my humor to use the Doctor's utterances and opinions against himself and his followers, much as it was Sampson's humor when he slew a multitude with the jaw-bone of an ass.

CHAPTER VIII.

FERMENTED VERSUS DISTILLED LIQUORS.

It is a common error to suppose that because alcohol is a constituent in both fermented and distilled liquors, that therefore the effects in the animal economy of both kinds of liquors are similar in their character, but differing in degree. It is by no means certain, however, that the alcohol of fermented liquors is in the same chemical condition as that existing in distilled liquors. There are many reasons which lead the chemist to believe that a chemical combination exists between the alcohol of fermented liquors and the various ethers and other substances with which it is brought into intimate connection. Loose this combination may be, and readily broken up by heat, but certain chemical reasons, not to mention physiological facts, tend to substantiate such an idea. Distilled spirits are almost entirely alcohol plus water; while wines and beer are alcohol plus water, plus ethers, plus nitrogenous matter, plus important salts, which greatly modify the effects of alcohol alone upon the system. "It would be a gross error," says one of our most recent authorities on the subject, "to conclude that a given quantity of wine is equivalent to an equal quantity of water containing a like percentage of alcohol."

The nearer we get to absolute alcohol (a substance only found in the laboratory of the scientific chemist) the greater becomes the danger of taking a narcotic dose by reason of its relatively smaller bulk. The danger also is incurred of injuring the delicate tissues and membranes of the digestive organs by the abstraction from them of water, by the concentrated alcohol. The former of these dangers is very considerably reduced in the case of fermented beverages; the latter, almost, if not entirely, removed. An instructive illustration of the difference in the effects produced by concentrated and diluted doses is seen in Bardeleben's experiment upon a dog. In this experiment it was found that forty-five grains of common salt, introduced at once into the stomach through an opening, occasioned a secretion of mucus, followed by vomitings; whereas, *five* times that amount of salt in *solution* produced neither of these effects. The explanation is simple. In one case the salt in order to pass into solution was forced to abstract from the tissues with which it came in contact the necessary amount of water. In the second case solution having been previously effected, the demand for water being supplied, the salt was practically inert.

Alcohol is the active principle of fermented liquors as theine is of tea, acetic acid of vinegar, or piperine of pepper. To use it as an ordinary beverage in the place of the primary fermented liquors, is as physiologically wrong as would be the substitution of theine for the common tea beverage, acetic acid in place of vinegar, or the oil of mustard for the ordinary condiment. Nature has placed a limit upon the strength of fermented liquors. This limit is practically placed at fourteen to fifteen per cent. of alcohol by weight. The moment alcohol is present in these proportions the yeast cells cease their activity, so that although much more matter capable of being broken up into carbon dioxide and alcohol may be present, further fermentation ceases to be practicable, indeed the addition of extraneous spirit beyond this fixed percentage results not only in checking their vitality, but in their utter destruction.

The only possible way then to obtain a liquid containing a high percentage of alcohol is by methods wholly foreign to and apart from the phenomena of fermentation. In the case of spirits this is attained by evaporating off the water at low temperature, or separation of the alcohol itself by distillation. When the active principles are separated thus from their primary natural compounds, they are removed from the category of ordinary foods or condiments and enter the province of medicine. For dietetic purposes in some cases distilled liquors are beyond doubt most valuable, but their use here approximates closely to that of medicines and in many such cases it would be well to obtain the advice of the physician, although when the individual is intelligent and fairly observant of himself "he is," says Dr. Bernays, "a better judge of what agrees with him than any physician can be."

Much of the harm which occurs from the abuse of alcoholics arises from the use of spirits as a beverage. The testimony in support of this view is so overwhelming in its character and quantity that it is amazing to see the conclusion so much as questioned. It is not a question of individuality here, but a question of whole races or nations. On the one hand we have the sobriety of the wine districts of France, Spain, Italy, and the Rhine. On the other hand, the spirit-drinking portions of north-eastern Germany, the manufacturing departments of north-eastern France, and certain cantons of Switzerland. In a most exhaustive scientific report recently published by the Federal Government of Switzerland, with statistics gathered and facts collected from all parts of the world, one of the chief factors in the promotion of intemperance was shewn to be the comparative dearness of wine and the impurity and relative cheapness of ardent spirits. The cheapness of wine then becomes not a prolific source of drunkenness, but, as noted by Adam Smith in his " Wealth of Nations," an aid to sobriety. Candid and careful observation will shew that in those parts of Europe where intemperance is rife, especially in the vicinity of manufacturing towns, there wine will be found much too expensive for general beverage purposes, while on the other hand ardent spirits often of the vilest description will be found in cheap profusion.

In a lengthy report on alcoholism, issued last July by Drs. Tourelot, Devoisins, Lailler, and Paul Chéron, this fact is amply illustrated. They shew that in the Departments of the Seine-Inférieure, the Somme, and Rouen, wine is almost unprocurable ; at least it is quite beyond the reach of the majority, while cheap spirits made from grain flow in abundance. A curious fact significant to imbibers of brandy was also brought out in their report, which was that more spurious brandy is consumed in the three departments named than the total quantity of genuine eau-de-vie manufactured in all France.

But there is no need to go beyond the borders of our own Province for a refutation of the silly idea that the milder beverages invariably lead to a craving for stronger ones. There was a time, not so very long ago, when whiskey was the only procurable stimulant. To-day the beverage most in demand is beer, and even this has of late been giving place to lager, and no doubt would have done so to much greater extent than it has were it not for the foolish policy that drives out by criminal enactments the bulky liquids to make way for the concentrated. In many of the rural districts whiskey may, it is true, even in those places not cursed by prohibition, be found the prevailing drink, but examination soon discloses the fact that this is directly attributable to remoteness from brewing centres and the consequent difficulty of procuring supplies.

In view of these facts it behooves sober, intelligent men to pause and reflect on the consequences attendant upon checking a natural tendency towards the use of milder drinks. Rightly or wrongly, it matters not which in. this connection, the demand for stimulants is implanted in the systems of the vast majority of the community. Are we practical men ? Let us recognize this fact, and instead of vainly buffeting against nature in an attempt to muzzle a natural tendency, aid its course in the direction of what tends to temperate habits.

Although lengthy reference to the capabilities of Ontario as a wine producing country would be out of place in this connection, even did our space permit, still the fact is worthy of note, that authorities upon the subject have pointed out, time and again, that Ontario, especially the southern counties, should rank by virtue of its climate second to no country as a wine producer. It is a fact worth pondering upon. We have here a great help to the true solution of the question of intemperance. Shall we accept it, or with blind fatuity turn to smuggled spirits, opium, and debauchery ?

CHAPTER IX.

SOME GENERAL ERRORS OF TEETOTALLERS.

That the demand for stimulant-narcotics is merely the result of an acquired habit, for which there is no valid assignable reason other than that which might exist in a diseased or vicious appetite, may be sufficiently probable to minds prejudiced by teetotallism, but will hardly prove a satisfactory explanation of what to all intelligent observers is an exhibition of a universal craving. "It is idle to urge," says Dr. Anstie, "that the subject of a carefully prepared experiment can be made to live in apparent health without the use of those substances vulgarly called 'narcotics' if the practical fact be that *nations* cannot and never have been able to do without them."

There is absolutely no period of history—there is absolutely no nation upon earth—in which indisputable evidence of their use may not be found. Von Bibra puts the matter roughly but plainly : " Coffee leaves are taken, in the form of infusion, by two millions of the world's inhabitants. Paraguay tea is taken by ten millions. Coca by as many. Chicory, either pure or mixed with coffee, by forty. Cacao by fifty. Haschish is eaten and smoked by three hundred millions. Opium by four hundred millions. Chinese tea is drunk by five hundred millions. Finally, all known nations of the world are addicted to the use of tobacco, either by smoking, chewing, or snuffing." Professor Johnson also shows that there is no considerable tract of the earth's surface without some special indigenous plant of which the natives readily avail themselves, not merely for medicinal purposes but for everyday use as a stimulant-narcotic.

Are we to believe that nature has here laid a pitfall for her children by the general planting of a false instinct ? Impossible. Nature, a beneficent parent, points the way : Science, her handmaid, follows with the reason ; nor is the reason in the case of stimulants far to seek. The ordinary life of man, and this is particularly true in the case of the higher races, is wholly comprised in highly complex and heterogenous actions, and highly complicated and heterogenous food—food requiring little expenditure of energy for assimilation, is requisite for its perfect maintenance. Doubtless, in the full possession of animal health, with the digestive and mental powers unimpaired, and with no special demand on the vital energies, stimulants, in the ordinary sense of the term, would be superfluous. But what might this mean ? No dishes of daintily cooked food ; no pepper, mustard, vinegar, sauce, beer, wine, tea, coffee, or the thousand and one stimulants to appetite and digestion we regard in the presen', and rightly, as real necessities. But the idea is absurd. How many physicians are there who can point out one such case in their whole practice? Such cases of ideal health are indeed rare.

Then as to the assertion of a "reaction" following the taking of a dose of alcohol, and that a renewal and increase in the quantity becomes a necessity. There is no more "reaction" following the ingestion of a proper quantity than follows the full and complete digestion of an ordinary meal. What has been termed the "reaction" has been proved by Dr. Anstie to be the *direct* effects of a "narcotic" quantity. As to the necessity of increasing the dose, this would be serious were there any foundation for the assertion, but Dr. Anstie has also shown that it is only those who have *habituated* themselves to the "narcotic" effects of excessive quantities who are obliged to increase the quantity. Do we not see every day a man enter a tavern for the purpose of quenching a natural thirst, or to relieve the feeling of fatigue consequent on a hard day's work. To such, the first glass of beer is most grateful, and he turns away, the natural appetite satisfied; but he meets a friend, and, in accordance with the much abused habit of treating, is asked to have a second glass. He refuses, the other presses, until at last, not to offend a frie id (?), he offends his palate and stomach by wriggling down the second or third glass. This disinclination often amounting to actual disgust, is the revolt, the "red danger-flag" of nature. This experience sometimes does not end with only the "second" or "third" glass, but may be prolonged, and not on one day only, or two, but for weeks—aye, months and years, it may be--until tired nature gives way, and the wretched man awakes to find that there is no longer a "danger-signal" displayed, but *unnatural* appetite created, which grows by what it feeds on.

If there were any truth in the dogma that moderatio. necessarily becomes excess, how are we to account for the habitual temperance of the Spanish or Italian peasant? Here the habit of partaking of the ordinary wines of the country as part and parcel of their daily food has been transmitted down from father to son for several generations. Surely if such a tendency is to be feared in the father it will manifest itself more strongly in the child? The idea is merely a bug-a-boo existing simply in the minds of many who have not had sufficient time or opportunity to investigate the question for themselves, or raised by designing persons for achieving certain ends. Lewes puts the thing very neatly in his "Physiology of Common Life:" "Men drink one or two cups of tea or coffee at breakfast with unvarying regularity for a whole lifetime, but who ever felt the necessity of gradually increasing the amount to three, four, or five cups? Yet we know what a stimulant tea is; we know that treble the amount of our daily consumption would soon produce paralysis. Why are we not irresistibly led to this fatal excess?"

Lately an attempt has been made to wrest insurance statistics to the cause and support of teetotallism. Within a few years a new "temperance" insurance company, at the head of which were men of prominence in public life and in prohibitionist circles, issued a prospectus giving a table of life expectations of "total abstainers" and "moderate drinkers." The table at once impressed many insurers, and was made to do excellent service by the agents of the company, the difference in the life expectations of the two classes seemed so enormous. The same table, in the same form, has also been made use of extensively in prohibitionist literature, and the figures given are correctly taken from the original source—a physician's calculations of many years ago. The difference between the original and the copy is simply in the substitution of the words "total abstainers" and "moderate drinkers" for the words "temperate persons" and "intemperate persons," which appear in the original at the heads of the columns of life expectations. It is unnecessary to ascribe dishonesty of intention to the promoters of the company. It is more charitable, and probably more correct, to impute the mistake to a careless acceptance of a garbled statistical estimate, but nevertheless there is much significance in the fact that leading minds of the teetotal world can so readily receive error because it bolsters up a pet theory.

The real facts are that we have yet no reliable evidence, nor are we likely to possess sufficient data from the statistics published by insurance companies upon which to base a scientific opinion either one way or the other. What little we do possess, however, leans to the side of temperance rather than to that of abstinence. Apart altogether from the inferences drawn on a superficial glance by persons unacquainted with all the facts connected with life assurance, from the mortality returns of new companies compared with old companies which have attained their maximum death rates, there are facts which seriously invalidate the claims put forward by teetotal advocates in regard to the alleged greater longevity of total abstainers. Not one of the so called "temperance" companies can compare in small percentage of the death claims to total risks with the leading insurance companies of this continent, none of which discriminate in favour of total abstainers as against moderate drinkers, and the vast body of whose policy-holders belong to the latter class. Tables and diagrams have been constructed on the basis of ignoring very important facts, and by making selected and unfair comparisons in favour of total abstinence. To compare a class drawn very largely from men whose mode of living is simple in other very important respects than their drinking habits, with a class embracing many men of sedentary habits, often given to late hours, dinners, and not merely abnormal habits of life, as well as abnormal undertakings and enterprise not usually found in total abstainers, would be manifestly unfair, and especially since agents, in their anxiety to obtain policies, secure the admission, against the intentions of all insurance companies, of men whose "moderate" drinking is in reality excess, and even of men who are unmistakeably intemperate. Even were moderation very much better than total abstinence in conducing to longevity there are other things connected with the present status of the two classes which might completely obscure the fact in insurance statistics, especially when it is considered whence the small totally abstaining class of the few temperance insurance companies of England are drawn, and the large classes from whence are drawn the vast numbers of those insured in the great insurance companies.

The general mortuary statistics again serve to still further discredit the statements of the teetotallers. It is claimed by them that the reason of the alleged much better returns in death rates of the total abstainers over those of the moderationists are due to their immunity from special diseases which are most prevalent amongst the intemperate classes. It is admitted on all sides that the diseases which are apt to be engendered or accelerated are in the main cirrhosis, gastritis, hepatitis, ulcer of stomach, diabetes, nephria (Bright's disease), and nephtritis (inflammation of kidneys). On the other hand the use of alcohol is commended by all medical authorities of repute in cases of zymotic diseases, phthisis, many kinds of heart disease, and bronchitis. In these latter alcohol is of incalculable benefit. Now to prevent cavil or carping let us place all deaths in the former class to the debit of alcoholism, although it is by no means admitted by the higher medical authorities that alcohol is an important predisposing cause of kidney or liver complaints. We will find upon even a cursory glance at the mortuary returns that the whole number of deaths in the former section bear but a very small proportion to the number of deaths due to the latter class of diseases. Where the former number their hundreds the latter rank their tens of thousands. In other words, where alcohol *might* have been guilty of causing or hastening the deaths of *hundreds*, those cases where the same agent would have been of service in *preserving* life carried off their *thousands*.

It is said that alcohol is a potent factor in the increase of insanity. This is in the majority of cases a confounding of cause and effect. Drunkenness is usually a symptom, not a cause. There are a greater number of religio-maniacs confined in asylums than dipso-maniacs. It is absurd to say drink brought these last to such a condition. If the writer were equally silly with those who reason thus in regard to drink he would say: "What a terrible thing religion is. See how it has destroyed the minds of these unfortunates."

Errors in eating are as likely to end in insanity as are errors in drinking. The effects of the former are too obscure, although as great, for other than the medical practitioner to observe. But the effects of overdrinking are apparent to all. If the temptation to excess is greater in the case of alcohol than in ordinary foods, it must not be lost sight of that the plainness of the warning is also much greater.

Business troubles, by deranging the digestive organs, form a prolific cause of mania. So, also, improper, poor, or badly cooked food contributes in no small degree to swell the number of inmates in our asylums. So generally is this fact recognized that the chief portion of the physician's care is now directed to the patient's dietary. So far from alcoholic beverages largely predisposing their users to insanity, the very reverse is most probably the case. In the case of the overtaxed business man, unable to spare energy requisite to digest sufficient food to meet the heavy drain on his powers, alcohol would here, by aiding the digestion of some solid nutriment, if not by its own ready assimilability, yield the force necessary to relieve the strain and prevent what might otherwise become the complete prostration and overthrow of body and mind. Is it not pitiful to reflect upon the numbers who but for a foolish, nay, wicked prejudice might have been saved from lunacy and untimely death by the moderate use of one of the most wholesome and safe of all stimulants, light bitter ale ?

The history of the medical testimony in favour of teetotallism is interesting. Forty years ago the teetotallers got up a manifesto signed by two thousand of the medical men in Great Britain, recommending total abstinence from all alcoholic liquors, as tending to greatly benefit the general health. Great amusement was created by the discovery, shortly after its issue, that the majority of its signatories had signed, at or about the same time, a document setting forth the peculiar merits of one of the Burton ales, and strongly recommending its general adoption by the public as a good and highly beneficial beverage. Not many years ago the teetotallers got up another document : remembering the history of the earlier one they were more cautious in wording it. There was nothing in it of total abstinence. It stated that many persons held greatly exaggerated notions as to the value of alcohol as an article of diet. It also inculcated great moderation in prescribing it for medicinal purposes. Sir G. Burrows, then President of the Royal College of Physicians, was the first to whom it was submitted for signature. He not only signed it, but was led into permitting his name to be used in obtaining other signatories. Armed with the card of Sir George Burrows, and taking care to call on consulting physicians at the busiest hour of the day when their offices were thronged, they succeeded in obtaining a number of other signatories, many of whom, however, afterwards stated that they had signed without inspection, trusting that what came from the president of the college would be all right. It fell into the hands of a well-known Metropolitan surgeon, who not only refused to sign, but sent a letter of remonstrance to Sir George Burrows, to the effect that he "did not see how it is possible ' immensely ' to exaggerate the value of alcohol as an article of diet. I believe it to be one which is simply indispensable for the whole of that large class of persons who, while they are subject to large expenditure of nervous force, are unable to digest more than a very moderate quantity of the dietetic equivalents of alcohol, in the various forms of fat and sugar. I am, myself, among the most moderate of drinkers, and on three separate occasions have endeavoured to become a total abstainer. Each time my health gave way in the attempt, which now for some years past I have not ventured to repeat; and my experience as a practitioner has taught me that many others are in a similar case."

After a lapse of ten days the President replied that the suggestions for alteration came too late for acceptance, as the document had passed through the hands of one hundred and fifty persons. He concluded with the following paragraph, which certainly concedes everything contended : " *I entirely agree with you in the opinion you express about alcohol as an article of diet. I think to a large class of persons * * * it is indispensable, and I know many remarkable cases in confirmation of your own experience of the attempt to abstain wholly from alcohol."*

One of the most ingenious of shuffles by which teetotallers seek to evade the natural consequences of their false position is the invention of the " unfermented" wine theory. The chemist can hardly have patience to treat such an absurdity seriously. We have seen in a former chapter that fermentation is a natural process which in the case of wine, is brought about by those minute unicellular organisms, which form the bloom upon the ripe grape. Crack but the skin ever so slightly and mingling at once through the grape juice these cells *immediately* set up fermentation, which in warm climates becomes active in one or two hours. Only the most difficult efforts can stay this process, and only complicated processes can preserve the juice for any length of time. These processes are essentially modern, being due to the researches and labours of M. Pasteur. Yet with the coolness born of ignorance one Reverend Doctor has no hesitation in suggesting that the method of preventing fermentation, termed in honor of its discoverer Pasteurization, was in common use in Judea in our Saviour's time. It is a curious fact, is it not, that while wine-*vats*, wine-*lees*, wine *presses* and all the peculiarities of fermentation is familiar language in the Scriptures, no mention or even hint is given of what a contributor to one of our papers has wittily termed the " *Great Grape Syrup Industry* " ? It is curious, is it not, that absolutely no mention should be made of *glass* bottles with ground glass stoppers, or, as suggested recently by a young lady of brilliant intellect, *self sealing* jars and *cane sugar* ? But the thing is childish.

Some of the methods of preparing the so-called " unfermented " wines, are revealed by chemical analysis. In the majority of instances poor, thin clarets are subjected to distillation to get rid of the alcohol, sugar is added to sweeten and thicken, and elderberry to restore the colour partially destroyed by boiling. In others, the juice is boiled with sugar to a syrup, and *salicylic acid* added to prevent fermentation. In not a few instances which fell under the writer's own observation, *cider*, sweetened and coloured, has been sold under the name of unfermented wine, and these samples contained a notable quantity of alcohol. These instances will give a fair idea of the sophisticated abominations that are sought to be forced upon the public under the cloak of religion and philanthropy.

It is not uncommon to hear clerical and lay agitators proclaim as a matter of convincing testimony "that the medical profession of this and that city has almost unanimously condemned the use of alcoholic beverages." Whatever value this counting of noses may have it would be well to critically examine these alleged condemnations. They are generally founded upon replies to circulars addressed by teetotal associations to the local medical profession, and containing questions put in a form calculated to mislead the general public. Of course, most of the physicians who accept the teetotal creed reply, but the vast majority of physicians, including most of the ablest medical men, treat the circular with indifference, or even contempt. The replies are grouped, noses are counted, and the statement published which at once is distorted into an endorsation of general total abstinence by the medical profession of the city, when the truth is that only the small minority, and many of these of little account in the profession, have approved of the teetotal position.

The experiments of Drs. Smith and Parkes are often glibly quoted—without, however, giving details—by teetotallers, as favoring their extreme views—(the official organ of a powerful church had the temerity to quote Moleschott as an authority for teetotalism). Although both these doctors were strongly prejudiced against the use of alcohol when they started their experiments, their scientific honor forced them to record the fact that their most careful investigation failed to discover the slightest harm resulting from the ingestion of moderate quantities of alcohol, while harm invariably resulted from excessive or improper use. Some of the physiological deductions from his experiments, given by Dr. Smith are indeed, extremely fallacious, inasmuch as they are based upon results following excessive doses, such, for instance, as twelve ounces of brandy,—an extremely poisonous and narcotic dose for the average man—but these do not invalidate the fact made known by his experiments—that *no* harm whatever resulted from moderation. The result of Dr. Parkes' labors during the Ashantee campaign indicated the fact that alcohol given in large doses, or *during*, or *immediately preceding the performance of duty* was injurious, but when moderate doses were given after the labors of the day were ended, *no* harm resulted, but decided benefit was invariably obtained, and that in the comparison between total abstainers, those who abused or improperly used alcohol, and those who temperately indulged at the close of the day, the result was decidedly in favor of the last.

That a moderate quantity of alcohol, preferably in the form of beer, is not only without capacity for harm in those accustomed to athletic or outdoor exercise, but is positively beneficial, is not only predicated by physiologists, but amply attested by the dietaries of the Oxford and Cambridge boat crews. These dietaries embody the results of half a century of the highest scientific observation and experiment. The fact that two pints of beer form part of the daily dietary of two of the finest bodies of athletes in the world is sufficient answer to the stupid outcries of prohibition zealots that beer is injurious.

Some of the other errors to which brief reference may here be made are the persistent quoting of Generals Wolseley and Gordon, and Mr. Gladstone and others as total abstainers. In England an Archdeacon actually quotes Edward Hanlan as a bright and shining example of teetotalism. The Graces, and other noted athletes have also been improperly pressed into the service of the teetotal cause. In regard to Weston, who certainly accomplished remarkable feats in pedestrianism, much error prevails. He accomplished the last portion of his famous hundred mile walk on small quantities of brandy and champagne administered in oft repeated doses. Nearly every one of his notable feats has been exceeded by other non-abstaining athletes, and his greatest record, 5,000 miles in 100 days, resting on Sundays, a task accomplished on water, has been surpassed by that of Fitzgerald, who accomplished 5,306 miles in 100 days and on beer.

In the training and performances of athletes much of course depends upon the individual condition of the athlete and the precise conditions which are to be secured, but there is no doubt that beer is found under many circumstances to be practically indispensable.

The experience of polar expeditions has also been misrepresented. Whatever the reason why caution should be exercised in regard to the use of alcohol during exposure to great cold, the fact remains that alcohol in small quantities is referred to by Greeley in connection with the last polar expedition as exceedingly useful.

A common mistake of advocates of general total abstinence is the assumption that alcoholic beverages are intoxicating in proportion to the quantity of alcohol they contain. For example, it is concluded that of a beer containing three per cent. of alcohol only twice as much must be taken as of beer of six per cent. alcoholic strength in order to produce the same effects. This is far from the truth. The rate of absorption from the stomach into the blood differs greatly with the degree of dilution of the alcohol; and the quantity in the blood at one time is also affected by the rate of oxidation or elimination of the alcohol in the circulation. Intoxication depends chiefly upon the quantity of alcohol in the circulation at one time, not the quantity in the stomach. Hence the fact that with a very large proportion of mankind it is practically impossible to become intoxicated with lager, however great the quantity consumed. The evils induced by excess in lager are generally of a different nature and arise from a different cause.

The partial explanation just given will enable the reader to perceive why many imagine that alcohol has first a stimulating and then a narcotic action, that the latter is "overstimulation," and that the depressing effects incident to narcotism are in lesser degree the sequel of moderate use. It is the quantity of alcohol in the circulation at one time, not the quantity taken into the stomach which chiefly governs the difference in conditions. The taking of a decidedly narcotic quantity of alcohol into the stomach is followed first, by stimulation, then as the quantity which has passed from the stomach into the blood increases to a certain quantity, narcotism commences. Where only a stimulating quantity of alcohol has been taken no trace of narcotic effects appear, nor is there manifested even in slight degree the so-called " re-action "—confounded with the depressent effects of narcotism—which in cause and phenomena is wholly unlike stimulation.

In regard to the heredity of drunkenness much misconception also prevails. Unquestionably there are many cases where a predisposition to intemperance is transmitted from a drunken parent to a child. But it might be of service to some to suggest that intemperance, whether in eating, drinking, work, or many other things in which excess is committed, may bring about a debilitated physical and mental condition which, through the mysterious action of the law of heredity, may leave offspring predisposed, through inherited weak or irritable systems, to drunkenness. Is it improbable that as strong a tendency to alcoholic intemperance may be transmitted from teetotal parents guilty of excesses in eating and otherwise as from drunken ancestors ?

The facts attested by the universal experience of observant men in all ages establish the conclusion that alcoholic beverages are commonly beneficial, while injurious in excess or under certain circumstances. In these respects alcohol is not at all an exception to the general rule. Nature emphatically endorses temperance in all things, and punishes intemperance. The fault, where harm is done, is generally a moral one on the part of the individual. A moral use of alcohol, as of other gifts of heaven, brings its own reward ; an immoral use its own punishment.

CHAPTER X.

GENERAL CONCLUSIONS.

In the ignorant physiological errors associated with teetotalism lie crystallized the erroneous theories of a past generation—theories long since laid to rest in the scientific world. But error dies hard, and so we find to-day in the sub-scientific world these ghosts of earlier speculations invoked and re-habilitated and marshalled to the front to do duty in support of a modern system of ethical error so glaringly defective that it could not have been able by itself to make much progress.

As we have seen in a previous chapter, the general acceptance of Liebig's chemico-physiological doctrine was due, not so much to the high rank in science of its author, as to the fact that it was published at the juncture when the theories of former generations were crumbling at their base, ere yet the foundations of the new bulwarks of modern science had been laid. The doctrine that the disorganization of highly organized matter liberated energy, together with the clear conception that that energy was present, although latent, in the complex compounds themselves was an enormous step in advance, and effectually pricked the fantastic bubbles of his predecessors and the visionary ideas of an Archæus and the spontaneous generation of energy, as well as the ingenious, but equally untenable, hypothesis of the Brunonian system. This doctrine, that the breaking up of complex bodies into more simple forms liberated the energy latent in the bodies themselves, appeared not only perfectly sound in Liebig's time, but is found upon examination to lie at the foundation of modern science. The separation of animal energy from any relation to such animal power as is implied in heat beclouded even his vision, and while yet within reach of the grandest generalization of science since the time of Newton, he failed to grasp the full force of his own proposition, and his celebrated distinction raised with such rare ingenuity between the carbon and nitrogenous compounds fell before the logic of experiments and analysis. But Liebig was by no means inclined to sustain the position that alcohol was of no service in the animal economy. On the contrary, he placed a high value upon alcohol, ranking it only second to fats as a respiratory material.

The first to apply the doctrine of the correlation of the physical forces to the explanation of vital phenomena was the late Dr. W. B. Carpenter. According to this doctrine such compounds of carbon and hydrogen as fat, sugar, starch and alcohol, are to be considered force producers. But the experiments and conclusions

of the three French chemists seemed to undermine this view in the
case of alcohol. It was before the masterly experiments of M. M.
Baudot and Schulinus, and Drs. Anstie, Thudichum and Dupre
had completely refuted the Frenchmen's error that Dr. Carpenter
wrote his contributions to the alcohol controversy. These contributions contained views which were almost immediately refuted to
the satisfaction of the scientific world, and with the progress of investigation on the subject Dr. Carpenter very materially altered his
views, his prejudice giving way, so that during the later years of
his life he indulged daily in a moderate quantity of wine.

It was also many years before the experiments of Messrs. Baudot
and Anstie that Prof E. L. Youmans wrote an article on intemperance for a New York journal, which, at the solicitation of teetotal
societies, he embodied in a pamphlet against the use of alcohol.
But in a recent review of his life by his sister, which appeared in
the *Popular Science Monthly*, the fact is brought out that (like Dr.
Carpenter) his opinions underwent a change, and he therefore allowed the book to pass out of print.

Yet we find the early and hasty opinion of these scientists embodied in a text-book on Hygiene, issued by the Ontario Department of Education mainly at the instigation of a small but organized and persistently active body of ignorant women, while their
change of view and the vast array of eminent medical authorities
of recent years in favour of the moderate use of alcoholic beverages
are entirely overlooked or carefully ignored.

It is true that for some time certain physiologists of repute
misled by the bold allegations of the French chemists, were more or
less inclined against the use of alcoholic beverages. This attitude,
however, was but temporary, and was maintained only till the full
force of the refutation of the Frenchmen's error became apparent
to the scientific world. But meanwhile mischief had been done.
The statement that alcohol possessed no alimentary value was
eagerly seized upon by prejudiced persons, and did much service
in the cause of compulsory abstinence, and by reason of its extensive currency amongst unscientific and ill-informed people it still
continues to obscure in the public mind the real attitude of science
on the question.

To claim that medical opinion is by a vast preponderance at
the present day in favour of total abstinence even as a general rule
for the community, may betoken skilful generalship in the cause of
teetotallism, but is nevertheless dishonest. Quite recently the
celebrated Dr. Bernays stated that " for every medical man of distinction who is in favour of total abstinence I would find twenty
men who would be against it." For the past twelve years, or, in
other words, since the completion of the researches so ably conducted by Baudot, Schulinus, Thudichum, Dupre and Anstie, no
work of repute in the scientific world has been published in support of general total abstinence, but, on the contrary, medical authorities, commencing with Pavy, Dupre, Thudichum, and Anstie,

4

and later, Sir James Paget, Garod, Bernay., Bennett, Sir W. Roberts, and Lauder Brunton, men of the highest sianding in medical
and physiological science at the present day, have supported the
cause of temperance against general total abstinence, and recognized in alcohol a decided blessing to the human race.   A
coloring of the earlier tendency remains in some minds, but even
that is rapidly disappearing.

That the true opinions of medical science on this question
have not yet found their way into general public thought should
cause no surprise when we reflect on the extent, activity and
methods of teetotal propagandism. A numerous press devoted
to the cause of teetotallism and thriving on the " inflammation "
it seeks to foment, garbles the statements of medical men, represents
every warning against misuse of alcohol given by eminent physicians as a declaration in favour of general abstinence, returns to
and reiterates with singular pertinacity the views of teetotal physicians of a generation ago, cites the names and quotes the opinions of every obscure doctor who ventilates his teetotal hobby in a
medical journal or local newspaper, or at a " temperance " tea-
meeting, and creates a wide notoriety for men who, were it not for
the pomposity of their declarations, would scarcely be known a hundred miles from home. Newspapers, of late the followers rather than
the leaders of popular opinions, fill their " temperance column "
with every waif and stray of teetotal misrepresentation, gathered
" regardless " from teetotal sources, with a view to tickling the
fancy of large and active organizations whose financial and political support is naturally a matter of more or less solicitude.

How little reliable are teetotal statements may be imagined when
we actually find Liebig and Moleschott, to say nothing of King
Chambers, and other high medical authorities, cited in prominent
religious and secular journals as opponents of the temperate use
of alcoholic drinks.

Nor should surprise be felt were even a large number of medical men of local standing to lean to the side of total abstinence.
Scientific progress but slowly affects some minds, and besides, as
Dr. Oliver Wendell Holmes has truly said, " medicine, professedly
founded on observation, is as sensitive to outside influences, political, religious, philosophical, imaginative, as is the barometer to
changes in atmospheric density."   So the iteration of the dogmas
of teetotallism, incorporated as they are in a widespread creed, has
its reflex action in medical thought ; and sentiment, not science, too
often colors the physician's mind. What might be predicated as
the result of adding to the bias engendered in many men, though the
creed of a powerful religious denomination, by organizing teetotal
medical societies among students, with a view to further prejudic-
ing their opinions, at the outset of their medical careers, on a question which, though demanding calm and impartial investigation,
has been so obscured by heated controversy that to quote Milner
Fothergill, " the pure, white light of truth can scarcely be seen clear

of the colored rays around it." An instance of the undignified length to which men will go in support of a hobby was recently afforded when a well-known city physician appeared before a medical society to address it in favour of total abstinence, armed with " medical evidence " in the shape of short clippings from the local secular press.

That a large number of medical men have not publicly protested against the misrepresentations of medical science and medical opinion is no matter for wonder. It is a difficult thing for the majority of practitioners to speak their mind freely upon this subject, their practice and public ambitions largely depending upon their standing in church and society. The emoluments of their profession, the daily food for themselves and families are ever present in their minds and too concrete to admit of any sacrifice in the cause of abstract scientific truth. How many, also, are there who satisfy their consciences when teetotal misstatements are brought to their notice with the sophism : " These exaggerations and misstatements are certainly very foolish and very much at variance with the teaching of science, but their propagators are working in a cause which has an element of good in it, and it might ' seem ' that we are opposing, not their errors, but their intentions, and if they do not accomplish all the good they seek, at least they may not succeed in doing any great harm." Against such sophistry we most earnestly protest. No permanent good has ever been achieved by falsehood, nor can be ; harm always does and always will ensue.

The time for silence is past. The influence of this false teaching has already accomplished a vast amount of evil. It has lent a weighty support to a moral error, which threatens not merely to lead to much physical harm by the deprecation of the good to be accomplished in very many instances by the temperate use of alcohol, but to grave injustice, to oppression, and to general demoralization of the tone of public and private morality.

Would the world be better without alcohol ? No. To believe otherwise would be to take no cognizance whatever of human nature. It is not enough in this connection to plead that many men ruin themselves by intemperance. There is not a single principle, not one good gift of God, that is not by someone made the occasion of sin. Not a few of these good gifts are the occasion of vast evils. Beyond doubt the relations of the sexes have been provocative of more crime and misery than strong drink, Were there no reason for the use of alcohol other than that properly used it affords pleasure to humanity, that alone should make us reflect before throwing it aside. The cares of the world are too numerous, its pleasures for the vast majority too limited to permit of rash or heedless sacrifice. But apart from this there is overwhelming evidence to prove that the temperate use of alcohol is not only conducive but absolutely essential to the health and capacity for work of countless thousands, and amongst these some of the best and most useful of the race. Are we then to abandon a source of much in-·nocent pleasure, and in pursuit of an *ignis fatuus* inflict a positive injury upon a large proportion of the community ? Are the beliefs of millions of temperate reasonable people that alcohol is of service to them to be disregarded in this discussion at the dicta of dogma-tists and zealots ? Are the opinions of the vast majority of physi-ologists to count for nothing in the question of temperance versus forced abstinence ? Is Christianity itself to be eclipsed by a moral error which in its nature and effects is simply appalling ?

To ignore facts is useless. The accumulated experience of countless generations that alcohol is one of the choicest blessings of heaven it would be most unscientific to ignore. That testimony, reaching down from the early dawn of history and attested in the customs of the most progressive people and the most useful of men, con·ʹ·utes as significant a fact as any in the realm of natural his-tory. The dicta of a few theorists affected by a passing sentiment cannot alter facts: they still remain, and physiology can only fol-low with the imperfect interpretation. Here experience, scripture and science are as one.

The evils that are associated with the use of alcohol may be slightly lessened as the explanations of science obtain currency. Appeals to reason, affection, morals, above all, to the motives of Christianity, are to be chiefly relied upon. The remedy for drunk-enness is a moral one. As Dr. Bernays says : " The children in our schools should be taught that the Kingdom of God is neither in meats nor in drinks ; that temperance does not merely apply to drink, and should proceed from right principles ; in fine, that tem-perance is better than abstinence, and that its influence is far greater." To turn to prohibition is to seek to abolish the good in order to extirpate the evil. It is, moreover, a weak and utterly ineffective expedient to obtain the latter result. It is born of fear —fear that man left to his own common sense and to spiritual aid would succumb to the temptations surrounding him. But to fly to weak human enactments as a protection against human weak-ness is about as effectual as the attempt to lift oneself over a hedge by one's own boot-straps.

Although we place our faith and hopes for any permanent change for the better so far as morality is concerned, mainly on the dissemination of sound Christian principle and the teaching of sci-entific truths, there are secondary methods by which we believe practical good can be accomplished. Legislation must not and cannot successfully ignore human nature, and there is some hope in it so far as it respects the truest instincts of that nature. En-couragement by legislation and otherwise of the substitution of the lighter fermented beverages for ardent spirits, the discountenancing of the much abused habit of indiscriminate treating, and the sub-stitution of the continental habit of table instead of bar in public places, the passage of a special and stringent adulteration Act with provision of the necessary machinery for its effectual enforcement, would approve themselves to universal public opinion, and we are assured would do much to mitigate the evils of intemperance occasioned largely by the use of spirits in the form of drams, and aggravated, it is to be feared, by reckless and criminal adulteration. But that such measures, however practicable and feasible they may be, will be forthcoming while the attention of our legislators continues to be forcibly directed to fighting the use, rather than the abuse, of alcohol, is, perhaps, too much to expect.

APPENDIX.

APPENDIX.

The following extracts are taken from the articles published in the Contemporary Review, and republished in pamphlet form, under the title of "The Alcohol Question." The extracts embody the opinion of ten of the principle contributors to the symposium.

DR. R. BRUDENELL CARTER, M.D., F.R.C.S.

—" If we come to enquire in what way this small dose (half a wine glassful of brandy or whiskey) exerts a beneficial action, we are at once met, on the part of many of the advocates of total abstinence, by the assertion that alcohol is not a food. I have no inclination for a controversy about words, but, if we may accept Johnson's definition of food as "anything which nourishes," I do not hesitate to say that the advocates of total abstinence are mistaken. I have recorded a case in which an old gentlemen took no other food for many months, and was kept, not only alive, but in moderate strength and comfort, and with no remarkable emaciation, upon alcoholic drinks alone. He liked variety and rang the changes upon champagne, old port, brandy, the strongest Burton ale, and other liquids, some of which contained a certain amount of sacharine matter, but not enough to maintain life as he maintained it. Cases of a similiar kind are recorded by the late Dr. Anstie and others ; and nothing is more certain than that people will live upon alcohol and water for long periods. The evidence by which this is proved seems to me altogether to outweigh the opinions of those who declare that alcohol is not food, on no better grounds than that they are unable to discover how it nourishes, or what transformation it undergoes within the body.

—" We may assure ourselves by common observation, that the moderate consumption of alcohol is useful to many persons, and that it does not produce, at least necessarily, or in any but exceptional cases, the dire effects which have been ascribed to it."

ALFRED B. GAROD, M.D., F.R.C.P., F.R.S.

—" The majority of adults can take a moderate quantity of alcohol in some form or other, not only with impunity, but often with advantage."

—As a beverage, alcohol should be taken in very moderate quantities, freely diluted, and usually at or after meals."

—" Many can abstain from their accustomed alcohol without any unpleasant results, and some with marked advantage ; but others, when they have ceased to take it for a little time, experience symptoms indicating that the nutrition of the system is not fully kept up."

—" Alcohol in the different combinations in which it exists in the various fermented liquors produces different effects upon the system, and discrimination is necessary in the selection of beverages by different individuals."

C. B. RADCLIFF, M.D., F.R.C.P.

(Formerly Lecturer on Matiera Materia, Westminster Hospital)

—" Alcohol, properly used is o   it service, partly in keeping up the animal heat by supplying easily ....dled *fuel* to the respiratory fire, partly in producing nerve-power by furnishing easily assimiliable *food* to nerve-tissue, and partly in *lessening the necessity for ordinary food by diminishing the waste of the system* which has to be repaired by food."

—I cannot help saying that he who chooses to urge the poor to forego the *proper* use of alcoholic drinks for the simple reason that semi-drunkenness. and drunkenness are, what they are indubitably, evils of incalcuable magnitude, is no less than culpable—I cannot use a milder term—in a high degree.   I know that these persons are actuated by the sincerist wish to do good to their fellow creatures, and that they are at worst no more than wrong headed ; but I cannot allow that goodness in the advocate for any particular cause is to be allowed to take the place of soundness in argument.   Good wrong-headed people, you must allow, are very dangerous people.

ALBERT J. BERNAYS, Ph.D.

(Analyst to the City of London, England.)

—" If alcohol slays its thousands, water has also its victims, and they are often the best of the race."

—" The experience of mankind is better than individual experience, and so for every medical man of distinction who is in favor of total abstinence, I would find twenty men who would be against it. And if a man is observant of himself and is temperate in all things, he is a better judge of what agrees with him, under ordinary circumstances than any physician can be.

—" Next to beer, the best form of alcoholic drink is wine. The public owe to Mr. Gladstone many a benefit, and, among his many services, the introduction of wines at a moderate price."

—"The demands of the town-life on the nervous system, in the mere struggle for existence, are sufficient reasons for recommending the moderate use of wine."

—" The principle I contend for is moderation rather than abstinence."

—" At the present day it is a common thing to meet a friend in very bad health, and you ask him the cause : often it is owing to some experiment in teetotalism."

—" The children in our schools should be taught that the kingdom of God is neither in meats or in drinks ; that temperance does not merely apply to drink, and should proceed from right principles ; in fine, that temperance is better than abstinence, and that its influence is far greater."

Sir W. W. GULL, Bart, M.D. F.R.C.P. D.C.L., F.R.S.

—" I prize alcohol and wine as medicines ; there are cases in which it would be dangerous to do without it. If opium were used instead, the results would probably be fatal. One of the Greek poets writes, " There is an equal use in wine and fire to the dwellers upon earth. In the northern regions you want more stimulant and fire, in the south less ; and again, more as age increases and vitality diminishes."

—" In advising a young man of sound health as to whether he ought to give up alcohol I should consider his calling. I am not sure that I should not advise an out-of-door man, doing a good deal work, a carter for instance, to take some beer, as a good form of food, containing sugar and vegetable extract and very little alcohol, but a very small piece of beefsteak would make up the materials."

—I do not think we should be prepared to say that, speaking of the labouring classes, everybody could go without beer, as a food of a light kind."

—" People will not listen to the temperance societies, because they carry their theories too far. I do not think that you can start with the idea that there is no use in alcohol and no good in wine."

SIR JAMES PAGET, BART, F.R.C.S., D.C.L., L.L.D., F.R.S.

—" On the question of national health and strength, I cannot doubt, on such evidence as we have that the habitual, moderate use of alcoholic drinks is generally beneficial, and that in the question raised between temperance and abstinence the verdict should be in favor of temperance."

—" We have no statistics, and are not within reach of any, for deciding the question between moderation and abstinence."

—" As for the opinions of the medical profession, they are by a vast majority, in favor of moderation. It may be admitted that, of late years the number of cases has increased in which habitual abstinence from alcohol has been deemed even better than habitual moderation. But, excluding those of children and young persons, the number of these cases is still very small, and few of them have been observed through a long course of years, so as to test the probable influence of a life-long habitual abstinence. Whatever weight, then, may be assigned to the balance of opinions among medical men, it certainly must be given in favor of moderation, not abstinence."

—"Then we have some deductions from physiological observations which are supposed to indicate a mischief in even habitual moderation. But some of these are really such, that if, in the place of "Alcohol" we were to read "common salt" we should be led to conclude, if it were not for the experience to the contrary, that we are destroying ourselves by the daily excessive use of a material which, in its excess, can alter the constitution of our blood, or the permeability or other properties of our tissues. And even the best of the physiological observations on alcohol do not touch the question between abstinence and moderation, more nearly than as suggesting some of the directions which further inquiries should take."

—"The beliefs of reasonable people are, doubtless, by a large majority favorable to moderation rather than abstinence, and this should not be regarded as of no weight in the discussion."

—"Thus, then from all the witnesses to the evils of intemperance we fail to get any clear evidence that there is mischief in moderation. Looking further we find in them certain indications that it is, on the whole, generally beneficial."

—"We may safely say that there is a natural disposition among adult men to drink ; a natural taste for alcoholic drinks whether for their cheering influence or for any other reason."

—"As to working power, whether bodily or mental, there can be no question that the advantage is on the side of those who use alcoholic drinks. And it is advantage of this kind which is most to be desired. Longevity is not the only or the best test of the value of the things on which we live. It may only be a long old age, or a course of years of idleness or dullness, useless alike to the individual and the race. That to be most desired is national power and will for good working and good thinking and a long duration of life fittest for these ; and facts show that these are more nearly attained by the people that drink alcohol than by those who do not."

—"I have dealt with the question between temperance and abstinence entirely from the side from which my profession has enabled me to study it so far as may justify my giving an opinion on it. My study makes me sure as I would ever venture to be on any such question, that there is not yet any evidence nearly sufficient to make it probable that a moderate use of alcoholic drinks is generally, or even to many persons, injurious ; and that there are sufficient reasons for believing that such an habitual use is, on the whole and generally, beneficial. It may be assumed that further study of the matter, by competent and calmly-minded scientific persons, will discover many facts concerning the use of alcohol which will lead to the remedy of such harm as, even in moderation, it may do to some persons, or to some whole races of men, and to its use being better directed and limited than in our present customs. But knowledge of this kind will not change the general conclusion in favor of the general utility of a moderate use of alcoholic drinks ; and till this knowledge is gained, everyone may assume that he may safely use them in such moderation as he does not find to be injurious.

—" But as I have said, there are many, who, even if they would admit this, would yet maintain that the mischiefs of intemperance are so much greater than any conceivable advantages of moderation, that we ought not to promote or defend moderation, because its promotion hinders the general adoption of total abstinence, which they say, is the necessary and only sure remedy for intemperance. Here I can only doubt. I should think that in this, as in other things lawful yet tempting to excess, the discipline of moderation is better than the discipline of abstinence."

—" But some will say what is this moderation ? How may we define it ? Let those who thus ask try to define, to the satisfaction of any ten persons, what, under all circumstances and to all people, is moderation in bread or the wearing of jewels, in hunting or the language of controversy."

T. LAUDER BRUNTON, M.D., F.R.C.P., F.R.S.

—" We will first consider what claims alcohol has to be reckoned as a food and perhaps this can be best done by comparing it with a substance, like sugar, whose claim to the title of food no one doubts. If we find that alcohol possesses those qualities which entitle sugar to rank as a food we must admit that it also deserves the name. Sugar disappears in the body as the fuel does in the steam engine : and although it will not support life if given alone, yet along with other food it will supply energy for increased work, or prevent the body from wasting. In these points alcohol resembles sugar. It disappears in the body, and although it will not of itself support life entirely, yet instances are on record of persons having lived for a considerable time with scarcely any other food. Hammond observed also, that when his diet was insufficient, the addition of a little alcohol to it, not only prevented him from losing weight as he had previously done, but converted this loss into positive gain. The objection may be urged that some observers have found alcohol pass out unchanged from the body and that it therefore cannot be ranked as a food. But the same objection applies to sugar for the experiments just referred to were made with large quantities of alcohol and when much sugar is taken at once, it will also be excreted unchanged."

WALTER MOXON, M.D., F.R.C.P.

—Speaking of teetotal societies he says :

" But the truth must be said that their success is deplorably small as estimated by the number of drunkards they reclaim.—The teetotal organizations show considerable apparent achievement when they turn to prevent the use of liquor by those who have shown no tendency to abuse it. But unhappily there is a drawback to this sort of gain."

—" I believe that to a large extent teetotalism lays foremost hold on those who are least likely to become drunkards, and are most likely to want at times the medicinal use of alcohol, sensitive, good-natured people, of weak constitution, to whom the sacred ecclesiast directed his strange sounding but needful advice. " Be not righteous over-much, neither make thyself over wise : why shouldst thou destroy thyself ? "

CHAS. MURCHISON, M.D., F.R.C.P., LL.D., F.R.S.

" Alcohol is useful in the course of most acute diseases, when the organs of circulation begin to fail, as they are apt to do. A moderate quantity usually suffices. The large quantities—*e.g.* one or two bottles of brandy in twenty four hours—still sometimes administered, may do harm by inducing congestion of various internal organs.

—" In convalesence from acute diseases, or from other weakening ailments, where the circulation remains feeble and the temperature is often subnormal, alcohol is also useful in promoting the circulation and assisting digestion."

—" In persons of advanced life the circulation is also often feeble, and a moderate allowance of alcohol often appears to be beneficial.

—" In all conditions of the system characterized by weakness of the circulation the daily use of a small quantity of alcohol is likely to be beneficial at all events for a time.

J. R. BENNETT, M.D., LL.D., F.R.S.

President Royal College of Physicians.

—" Has alcohol any special advantage over other articles of diet in restoring exhausted nervous power or repairing the waste attendant on its exercise ? I believe it has, and that where one man may be met with who finds " a few raisins " answer the purpose, there are men whose experience has told them that " three or four brandied cherries " are better, and the majority of those who have to go through the labors of a parliamentary session or any similiar continuous mental strain, will I am convinced, admit that they do their work better and with more comfort to themselves if they take three or four glasses of sherry or claret as a part of their daily food."

—" Judging from my own experience both personal and professional, there is need for every one to relax his rule and modify his practice according to the varying circumstances of his life ; and to this most men's instincts prompt them. Many a barrister or a doctor in his summer holidays feels that he does not need his customary glass of sherry or port, does not care for it, and does not take it; but he no sooner returns to his duties than he becomes conscious that he is happier, more comfortable, and ready for his work by resuming his accustomed habit. I do not believe that such a one is, *cæteris paribus* a worse, but a better life for an insurance office than a pledged abstainer."

—" If every man is to forego his freedom of action because men make a licentious use of it, I know not what is the value of my freedom. If in the case of alcohol, as of meat, or any other thing, I am to abstain from what I conscientiously believe to be the lawful and beneficial use of it. " Lest I make my brother to offend," my life would be an intolerable burden, worse than that of any ascetic monk that ever lived, and, worse still, I should be perpetually giving the lie to what I believe to be a truth, that "every creature of God is good, and to be received with thanksgiving."

The following scientists, amongst the highest in America, were examined before a joint special committee of the Legislature of Massachusetts at Boston in 1867 : Professors Louis Agassiz ; Edward H. Clarke, M.D. ; Oliver Wendall Holmes, M.D. ; Henry J. Bigelow, M.D. ; N. Horsford ; J. B. Jackson, M.D. ; Charles T. Jackson, M.D.

The answers given to the different questions put to them, will be found interesting reading in view of the fact that their testimony was given at a time shortly after the publication of Dr. Carpenter's work on alcohol ; and when the experiments of Lallemand, Perrin and Duroy together with their complete refutation and exposure by Drs. Anstie, Dupre and Baudot were still fresh in the minds of physiologists and chemists.

PROFESSOR LOUIS AGASSIZ.

The Renowned Naturalist.

—" In the liquor drinking part of Europe we find intemperance, but intemperance is unknown in the wine-growing countries. It was when I went to England, for the first time in 1834, that I lei ned what drunkenness was In England and in Northern Europe intemperance is at home ; but is unknown in other parts of Europe where the grape is grown. I hail with joy—for I am a temperance man, and friend of temperance—I hail with joy the efforts that are being made to raise wine in this country, and I wish them all success. I believe that when you can have everywhere cheap, pure, unadulterated wine, that you will no longer have need for either prohibitory or license laws."

—" If I understand you correctly, the use of such wines does not engender any morbid appetite, resulting in drunkenness ? A.—I have never seen any morbid appetite engendered by the use of pure wine, any more than the using of other food engenders a morbid appetite for more food, or for food that is injurious."

- Q. " Your attention has been called to the effect produced upon those who come from the old country to this by prohibitory legislation ?"

A. " I think that the effect is an injurious one ; and I perceived very early in my residence in the United States, that there were a great many people who shared my opinion and my feeling about the use of pure wine, but who were led by the pressure of the sentiment of the community to publicly express opinions very different from those they really entertained, and who practised in private what they denounced in public. I have observed so much duplicity in regard to the subject, so much difference between public profession and private practice, that the whole question has become so unpleasant to me that I have generally abstained from making enquiries."

—" Not long ago, a clergyman of the highest respectability told me that he could not perform his duties without sustaining his system by an occasional glass of wine ; and he added that " such was the prejudice of the country that he dare not let it be known for fear of losing his influence."

DR. JAMES C. WHITE.

Professor Chemistry in Harvard University.

Referring to the experiments of the three French chemists, Messrs. Lallemand, Perrin and Duroy.

—" These experiments have been repeated at a much more recent day, and by men occupying as high a scientific rank, and the result of all the experiments is precisely the same. But the conclusions to which these other chemists have come are quite different. That is, they find a small percentage of alcohol excreted, but they say any conclusion as to quantity, based upon that amount is unfounded ; that they cannot assert, from finding a small amount, that ALL is excreted. And this is the opinion of ALL SCIENTIFIC MEN that I know at the present day. Moreover experiments have been made by other chemists, by taking the whole amount of the excretions passing within twenty-four hours and with very much larger, and with very much smaller amounts than these French chemists used. It was found by them, that in the case of a person taking twenty-four ounces of brandy, from time to time, during twenty-four hours, the amount of alcohol discharged by the kidneys (which is the chief channel according to these chemists) was very small indeed. The amount of fluid, containing enough alcohol to burn which they obtained was a very few drachms. So the conclusions of the French chemists were certainly unwarranted."

—Q. " Then what, if anything, have you to say in reference to the final conclusion of those French chemists, for the reasons they stated alcohol was not to be classed as a food ? "

A. " That they offered no evidence whatever."

.—Q. There is a theory, is there not, among physiologists which reckons alcoholics as in the category of foods ? "

A. " There is evidence, from their physiological action, that under some circumstances, they act as food, in the same way, for instance, as beef-tea does ; their effects are precisely the same as food, judging by their effects alone."

—Q. " Is, or is not the theory stated by you—as to the theory of alcohol being food—generally accepted by physiologists ? "

A. " I think that it is."

DR. J. B. S. JACKSON, M.D.,

Professor Morbid Anatomy, Harvard.

—" I refer, speaking of the diseases that seemed to be attributable in a few cases to the use of liquor, to certain diseases of the kidneys which have only been observed in modern times, and also to diseases of the liver. Both of these are common diseases. But I would also say, that both of these diseases I have often met with anatomically in persons perfectly temperate. But on the other hand, I refer to turbercular disease, which is so common, that it is said that, in the temperate regions, it destroys about one out of every five or six in the community, as it has been observed in various countries. This is the result in Europe, and in the observations which I have made I came to the same conclusion—about one in five and a half. I had occasion several weeks ago to refer to these observations of my own, so that I can speak definitely, and in these observations I noticed a fact I had not thought of before. It was not till I had proceeded some way that I noticed that quite a number of the grossely intemperate subjects that I examined had perfectly healthy lungs. From that time, I went on and examined all the cases with this point in view, to ascertain whether those who had perfectly healthy lungs had been temperate : and in regard to those whom I knew to be grossely intemperate, whether they died of consumption, and whether they showed that there had been formerly tubercular disease. And I arrived at the conclusion that I just now stated, viz.:—i have met with diseases, not unfrequently occuring in persons who had used liquor to an inordinate extent, which I suppose I had reason to think was owing to such use. And I was struck with the fact at the time that I made the examinations to which I refer, that a very important class of diseases seemed to be rather unfrequent in those who had been addicted to the use of their liquors."

PROFESSOR E. N. HORSFORD, M.D.

The well-known Analytical and practical Chemist, for sometime a student with Baron Liebig.

—" Alcohol comes under the head of what is called respiratory food, which includes starch and oil and those substances which when burned keep the body warm. "

Q. " Have you any opinion whether any form of alcoholic drink does act as food ? "

A. " I have, sir. "

Q. " What is your opinion ? "

A. " I think it is a food."

Q. " Will you explain ? "

A. " It ministers to the strength of the organism, and it may also enter the organism. It is allied to fat, and substances which produce fat, and in so far as it renders more perfect the digestion of food, meat and farinaceous food, it acts itself as food. Perhaps the most recent experiment that has been performed, is an experiment going to show that all these classes of bodies do actually fulfil the office of food, and that they do enable a man to perform feats of strength, which he could not otherwise do. These experiments have been tried in Switzerland."

—Q. " Do you make an essential and characteristic difference between a large dose, and a small one ? That is, does a small dose, a mere stimulating dose, have an effect characteristically like a large or narcotic dose ? Or is the effect characteristically different ? "

A. "There are two stages of effect, one of a stimulant, and the other of a narcotic character, both of which have their office. If you mean to ask whether, because it is a poison in its absolute purity, it is absolute poisonous when in combination with other substances ; because there is acetic poison in vinegar, it does not therefore follow that we may not use vinegar ; or because there is poison in the oil of pepper, that therefore we cannot use pepper. The essential principle of mustard and horseradish and a large number of condiments, which in their dilution are not poisonous, act as a poison on the human system, and yet the condiments themselves are very commonly used, and are found to be beneficial to a certain extent."

—Q. " If the question was asked you, whether God made or produced alcohol, what should you say ?

A.—" I should say he did. It is something which is spontaneously produced at times, and does not require the operation of human agency in order that it may be produced."

PROFESSOR EDWARD H. CLARK, M.D.,

Professor Materia Medica, Harvard University

—A. " I have lived at one time for about three years in an almost exclusively vine-growing country, and I looked upon the light wine there produced as being an addition to the comfort and sustenance of the people. I saw but very little drunkenness."

—Q. "You recognize, then, if I understand you, that there is both a scientific, and a practical difference between a stimulant dose and a narcotic dose ?"

A.—"We do. There is both a double, and treble action of many liquids, and alcohol is no exception ; in certain doses it produces an effect which is characteristically different from the effect produced by any other dose."

—Q. " Can you or can you not, in all cases, substitute what is properly called food, and thus get along without these stimulants ? "

A.—" You cannot ; the process (of life) sometimes will not go on without something to aid it ; or it will go on imperfectly, if at all."

—Q. " Do you not frequently find persons in a diseased condition, induced by the unnecessary use of alcoholic beverages ? "

A.—" Undoubtedly, but I should say, so far as my personal experience goes, not a deal oftener in this city (Boston) than I do from the taking of opium."

—Q. " Is it demonstrable that the nutriment of these preparations (Alcoholic) cannot be administered in any other form ? "

A.—" I know nothing that will take the place of them."

Q.—" You speak of diseases now, do you not ? "

A.—" Of disease in a general way. I speak of people who are well enough to be about their ordinary business and are called ' well.' "

—Q. " What do you think would be the result upon the general health of the people, if the whole community should determine no longer to take their aliment in the form of alcoholic preparations, but in some other way ? "

A.—" I think they would seek some other form of excitement. I can answer that question best by an illustration—at the close of one of my lectures upon opium, one of the class, who was formerly a druggist, from an interior town in the western portion of Massachusetts, came to my room and said he believed that in the town in which he lived the prohibitory liquor law was perfectly enforced. The town was so situated (being remote from a railroad) that there was no opportunity for dissipation, and but little opportunity for obtaining liquor. As far as he knew ; that law was enforced. He stated that he had been a druggist in that town for many years ; that previous to the enforcement of the prohibitory law he sold opium only on the direction of a physician ; since then he had sold for the last two or three years an average of one grain per day of opium to every man, woman, and child, in that town. The people had sought another means of excitement, another stimulant."

—" There are very few persons in the community in perfect health, probably not one of whom you could say with absolute certainty that every part of his body was in a state of perfect health, but for average individuals, who are in average health, you will find that with some, the digestion is imperfect ; the food cannot be converted into bone and flesh and brain, because there is an insufficient power in the machinery to work it up into that material ; there are certain conditions of that sort, where the addition of some kind of stimulant—which might in one case be one kind, and in another, another kind—will enable that process to go on very perfectly, perhaps as perfectly as it does in the average of cases. Under those circumstances, a person may be about his ordinary work, seemingly very well, and yet requiring as an aid to his digestion, some agent which will enable the process of alimentation to go on, and it will go on much better with, than without it. Or a person may be troubled with some sort of disease by which he is wasting away more rapidly than the system ought to waste, so that the building up of the system is not equal to the destruction of it, and such a person becomes emaciated, and yet in a condition of fair and average health ; but such wasting away will tell upon him sooner or later. Under such circumstances, the use, AS A PART OF FOOD, of some agent like that we have described will arrest the destruction, aid the process by which the system is built up, and in this way contribute to promote the health. There are a great number of cases of that kind where the dietetic use of alcoholic liquor is so important that I have deemed it necessary to give a lecture to the class upon the proper dietetic use of this class of agents."

Q. " Are there any cases, that will admit persons who are properly regarded as in health, and who are pursuing their regular avocations year after year, and making up a good average life by that sort of ailmentation, who otherwise would sink beneath the cares and burdens of daily duty and life."

A. " I meant to include what I suppose you refer to, under what have just stated."

DR. OLIVER WENDELL HOLMES, M.D.

Parkman Professor of Anatomy and Physiology.

—" I think, that alcohol, sir, is meant for use in the arts, for burning in lamps, and for various other similar purposes. Alcohol itself, I do not think, is an article which is used in the human economy. I have never known it to be drank unless some person may have been unfortunate enough to have drank it from the jars containing specimens in anatomical museums."

Q.—" That question lays the foundation for the one next in order, which is this : have alcoholic drinks, or drinks into which alcohol enters as a consituent in combination any proper use in the human economy, dietetically or medicinally ?

A.—" I think they have, both dietetically and medicinally."

Q.—" In what way do they act dietetically ? "
A.—" They act as food—these combinations."
Q.—" And also as medicines ? "
A.—" Of course."

PROFESSOR HENRY J. BIGELOW, M.D.

—Q. " You have travelled pretty extensively on the continent ? "
A—" Yes, Sir. In England, France, Italy, Egypt, Germany, Switzerland and other places."
Q.—" Have you any opinions as to the effect produced upon the health of the people of the wine-growing countries, by the use of wine as a DAILY BEVERAGE ? "
A.—" I should judge from observation of people as you see them, and from the entire absense of evidence of injurious effects, that it was not injurious in any way."
—Q.—" What did you observe in your several visits in reference to the sobriety and the habits of intoxication of the people there, either one way or the other."
A.—" I should say there was a very marked absence of ything like intoxication as you see it upon the surface of society."
Q.—" Do you say that in regard to those people who use wine daily as a part of their daily food ? "
A.—" I speak of those."
—Q.—Is it or is it not, your opinion that the wine they used did perform the office of food ? "
A.—" I have no doubt that it did to a certain extent."
Q.—" Did you attribute a good effect as food, to the alcohol or the other ingredients ? "
A.—" To the alcohol in combination with the other ingredients ? "
—Q. " In relation to the influence of wine upon the general health, you are able to state whether the long-continued use of wine results in any common class of diseases, insanity for example, in any considerable number of cases ? Say, take it in France. ? "
A.—" Not to my knowledge, when taken temperately."
Q.—" By temperately, you are speaking of its use habitually with the meals ? "
A.—" Habitually with the meals ; and based upon the fact that the light-wines are constantly drank, year in and year out, in almost any community in France. If alcohol were used to produce a stimulating effect alone, I should not expect diseases ; but if it were used to excess so as to produce a narcotic effect, I should expect diseases to result."
Q.—" Are these different stages of the same cause ? "
A.—" I incline to the belief that there is a characteristic diffference. So that you can put your finger upon the interval, and say that one is a stimulant effect, and that the other is a narcotic effect.
—Q. " Do you believe that the ordinary usages of society require the use of stimulants ? "
A.—" Undoubtedly, like all habits, this habit is liable to excess."
Q.—" Is that the strongest remark you would make ? "
A.—" I should say that excess is not a good thing, but, for a little excess you will find a vast amount of wine drinking, and the stimulus, on the whole, to the advantage of the individual."
—" Q.—" On the whole, you would say that the drinking usages of the community about us are to be reprobated or deplored ? "
A.—" Deplored ? No Sir."

PROFESSOR CHARLES T. JACKSON, M.D.

Former Professor of Chemistry, Harvard University.

—" Alcohol and alcoholic liquors act as respiratory foods. When alcohol either in the form of wine or in the form of any of our distilled spirits sufficiently diluted with water, is drank, it goes into the circulation. The alcohol takes the place of so much food in our bodies in the process of respiration. The carbon and the hydrogen are oxidized and removed from the system, and thus save the consumption of just so much tissue.

The moderate use of alcoholic drinks, so far from doing harm to the human body, serves to sustain its power of endurance, and saves the destruction of so much of our tissues, and is, therefore, conservative to the system."

—In answer to the question of whether the question of alcohol serving as food is unresolved, he answered :

" I do not consider it unresolved. It is not so considered by scientific men generally. There may be some doubts raised by some persons ; but I think that the opinion of scientific men generally is the same on that point."

STIMULANTS AND NARCOTICS.

By Francis E. Anstie, M.D., M.R.C.P.

Lecturer Materia Medica, and formerly Lecturer on Toxicology, Westminster Hospital.

—" Common salt is, in small doses, a perfectly indispensible article of human food, without which we should perish miserably ; in medium doses, it is a safe and useful emetic medicine ; while in extremely large doses, it is an irritant poison, and has caused death in several cases."

—" It is idle to urge that the subject of a carefully prepared experiment can be made to live in apparent health without the use of any of the substances vulgarly called Narcotics, if the practical fact be that nations cannot, and never have been able to do without them."

—" One thing is specially to be noticed—that all really irritant action (i.e. which is capable, if prolonged, of causing inflammation), is of a radically distinct, if not opposite kind from whatever increases the proper function of a part; whatever, in fact makes it to be more alive. We have no right, for instance, to speak of the agglomeration of blood copuscles which takes place when mustard is applied to a delicate web of nearly naked capillary vessels, as a higher degree, or as any degree, of the same action which is produced when we swallow minute quantities of mustard with our food."

—" If stimulation means the calling forth, that is, the getting rid of a certain quantity of a force already existing in the organism, either the accumulated stores of this force must be immense, or they must be simultaneously repaired by that which can create force, or the vital power must after a very short time become completely exhausted, and the patient, whether cured of his fever, or not, must be "improved off' the face of the earth." The simple facts of the case, however are these : a fever patient often lies for days taking an entirely insufficient quantity of common food to create new force in place of that so lavishly wasted by the stimulant medicines, possibly taking no such food at all. Yet at the end of a certain time, it may be many days, it is discovered that the necessity for the stimulant medicine has ceased or diminished, not by the patient's death, but by his having recovered his energy so far as to digest common food. Nor is it possible to get over this difficulty by supposing that the tissues have been consumed to such an extent that force has been generated at the expense of bodily bulk ; for the patients are comparatively very slightly emaciated on convalescence, if they have been treated in this way. Has there been great emaciation, so to speak, of the *nervous system*, though this has been invisible to our eyes ? It does not appear so, for the patients recover intelligence, sensation, and voluntary movement with great rapidity. In short, on the theory that stimulant action is followed by a more than equivalent recoil, we are entirely unable to explain these facts. But if we agree that stimulant action is, indeed, followed by a recoil, but that the latter is not greater, on not so great, as the original exaltation—the statement loses all importance, since this is exactly what happens AFTER THE DIGESTION OF A TRUE FOOD."

— "Alcohol taken alone, or with the addition only of a small quantity of water, will prolong life greatly beyond the period at which it must cease if no nourishment, or water only had been given : that in acute diseases. it has repeatedly supported not only life, but even the bulk of the body during many days of abstinence from common foods : and that in a few instances. persons have supported themselves almost wholly on alcohol and inconsiderable quantities of water *for years* If these things can be proved, as I hereafter shall show they can be, there is no need, of course, to argue further about the alimentary character of alcohol. We may be at a loss to explain the *chemistry* of its action on the body, but we may very safely say that it acts as a food."

— "Much might be said, and very forcibly, in refutation of the assertion, that there is a logical *necessity* for stimulation to be followed by a recoil equal in extent to the first elevation ; but we may save ourselves all this trouble by declaring simply that, in fact ; *no such recoil occurs*. The *narcotic* dose of alcohol. as will hereafter be shown, is alone responsible for the symptons of depressive reaction."

— " What depression is there as an after consequence, of a glass or two of wine taken after dinner, or of a glass of beer taken at lunch by a healthy man ? What reaction from a teaspoonful of sal-volatile swallowed by a person who feels somewhat faint ? What recoil from the stimulus of heat applied in a hot bath, or of oxygen administered by Marshall Hall's process. to a half-drowned man ? ABSOLUTELY NONE WHATEVER. The visible immediate results of these measures do, indeed, after a time, disappear, not being exempt from the ordinary condition of temporal things ; so that, just as food requires from time to time to be renewed, so does the oxygen which has been artificially driven into the drowned man's thorax require to be renewed by his own respiratory efforts, when he has once recovered the power to make any : and so does the glass of wine which we took to-day to relieve our sense of fatigue require to be repeated to-morrow, when similiar circumstances present themselves, as on that occasion."

— " I submit that it is more rational to regard the stimulant class of remedies as exerting a food action."

ON THE USE OF WINES IN HEALTH AND DISEASE.

By Dr. Anstie. M R.C.P.

— " It is amongst the class of natural wines, averaging not more than ten per cent of absolute alcohol, that we must seek the type of a universal beverage for every day life. If we turn to the most recent analysis (the very careful work of Dr. A. Dupre, of the Westminster Hospital), we find two kinds of wine which, as far as alcoholic strength is concerned, meet the ideal want—viz : Rhine wine at 9½ per cent alcohol, and a claret at 8½ per cent. Such wines are easily procurable, and we may say that we have in either of them a beverage which, alone, or diluted with a certain amount of water, would at once satisfy all needs for liquids with the principal meals, and all needs for alcohol, in the most convenient and agreeable way. A bottle a day of either of these wines for an actively employed adult, and a proportionaly less quantity for those whose life is more sedentary, would very well represent the allowance of alcohol which may be said to suit best the standard of ordinary health.

—We maintain that for the hard working student, politician, professional man, or busy merchant there is no better arrangement possible than that of taking, as the regular daily allowance, a bottle of sound ordinary wine of Bordeaux : and that the number of persons with whom such a diet really disagrees is very limited ; but on the latter point we shall have more to say hereafter, in discussing the other ingredients of wines. It may be added that no other wines which the world produces are capable of yielding, day after day, such unwearied pleasure to the palate as the sound ordinary wines of Bordeaux and the Rhine.

—"We did not intend, when recommending the "hard-working Student" to allow himself a bottle per diem of weak Bordeaux wine, to give that recommendation to young lads. We were thinking of "hard-working Students" of middle age ; and we would state our very firm conviction, that for youths (say under twenty five) whose bodily frame is as yet not fully consolidated, the proper rule is, either no alcohol or very little indeed.

Still, there can be no question that to many rapidly growing lads an amount of alcohol (preferably as beer) strictly limited to these latter quantities, is not only harmless but most actively useful."

THE MANAGEMENT OF HEALTH.

By JAMES BAIRD, B.A.

Weale's Scientific Series.

—"The habitual use of ardent spirits is altogether to be deprecated, even by non-teetotalers. When used, it should be in some emergency, when they may serve a temporary purpose, but there the use of them should cease. There is, and can be, no valid objection against the moderate use of pure or light wines, nor yet against wholesome beers. They do no tempt to excess, and they refresh and exhilarate. Brandied wines are open to the same objections as ardent spirits in a rather less degree. They should be used but sparingly."

WEALTH OF NATIONS.

By ADAM SMITH.

—"The cheapness of wine seems to be a great cause not of drunkenness but of sobriety. The inhabitants of the wine countries are in general the soberest people in Europe ; witness the Spaniards, the Italians and the inhabitants of the southern provinces of France."

Lecture delivered before the Exhibition Committee, London, 1884.

DIET IN RELATION TO HEALTH AND WORK.

By Alexander Winter Blyth, M.R.C.S., F.C.S, Etc.

— "There is a moderate and immoderate use of all things. Certain people acquire a habit of drinking enormous quantities of tea or coffee ; the majority, like smokers of tobacco, are not preceptably affected by the habit ; but a few fall into a dyspeptic nervous state. evidently due to excessive tea drinking."

WATER AND WATER SUPPLIES AND UNFER-
MENTED BEVERAGES.

Professor John Attfield, Ph.D., F.R.S.

—"Tea drinking to excess is only less harmful than alcoholic drunkenness."

SALT AND OTHER CONDIMENTS.

By T. T. Manley, M.A.

—Speaking of condiments, he says, " the golden rule of moderation —*ne quid nimis* should be observed in their use."

.—Speaking of salt in certain districts of Africa, he says. " Men will barter gold for it where it is scarce, and for it husbands will sell their wives, and parents their children."

—" The full title of Dr. Howard's work is ' Salt, the Forbidden Fruit or Food ; and the chief cause of diseases of the body and mind of man, and of animals : as taught by the ancient Egyptian priests and wise men, and by Scripture ; in accordance with the author's experience of many years." It is not too much to say that the writer is a painful instance as to the length people will go in perverting historical and other facts, and texts of scripture when they have once become the subjects of a craze, and dubbed themselves apostles of a movement.   Unfortunately such extremities will ever be found and the present age seems particularly prolific of them, whether the crusade be against the use of animal food, of tobacco, or of alcohol, or against vaccination.   With such it is of little avail to adduce facts or arguments, they admit nothing which tells against their views : they contort and pervert everything ; and the ' Counterblast ' of King James is their only type of reasoning.

—" In this matter of salt some few are indeed more temperate, and state their views more moderately and intelligibly. Thus, for instance, one of the opponents of salt eating puts the case, though with no little misconception of physiological facts. ' We have' says he ' among us physiologists of no mean standing,* who regard mineral salt as a poison, and as the predisposing cause, therefore, of many ailments. It is true that in its passage through the system salt remains unchanged, but this is only to say that it exerts upon the substances with which it is mingled in the system the antiseptic qualities that it does outside the system. That is, to a greater or less extent, it prevents the food from decomposing, and therefore from turning into nourishment. It is for this reason that salt is, as it notoriously is, a provocative of scurvy—a complaint which consists mainly in poverty of blood, through lack of nourishment owing to the action of salt. It may be granted that a diet of flesh requires a certain small amount of salt to retard the process of digestion and assimilation which otherwise would be too rapid, flesh not being the natural diet of man, who, in all physiological respects, is a grain and fruit-eating animal. But with our natural diet salt is worse than superfluous, saving only for the fact that we have become so accustomed to it that we do not like to do without it." '

—" That such views are held by a certain number of persons in this country may be admitted, and recently there seems to have been an attempt to organize something in the way of an Anti-salt association, the members of which were to abstain from salt themselves, and to endeavor to gain converts to their opinions and practice. Some years ago there was established in New York a Medical College, the prime tenet of which was abstention from salt."

PRACTICAL DIETETICS.

Professor Francis de Chaumont, M.D., F.R.S.

—" I tried with a friend to see how long we could live on this essence of meat, taken instead of ordinary albuminate food, on the supposition that it was equally nutritious and after a very few days we were reduced to a state of considerable inanition, and *exceedingly bad temper*, which was immediately improved by the addition of a little more proper food, especially a little butter."

ÆSTHETICAL USE OF WINE.

J. L. W. Thudichum, M.D., F.R.C.P.

—" Of all alcoholic drinks true wine, as we shall define it offers the least opportunity or inducement to abuse. Natural wine may make drunk, but it never produces *delirium tremens*, it never produces those permanent lesions of the tissues which are the consequences of excess in the use of spirits and beer : whenever such effects are added as the result of wine, they will be found to be due to wine *plus spirits added thereto*, particularly to those fiery mixtures which under the name of sherry and port have done so much to obscure the real and beneficial qualities of wine."

— 18 —

R. A. WITTHAUS, A.M., M.D.

Professor Chemistry and Toxicology, University of Vermont, and of Physiological Chemistry New York University.

—Referring to champagnes and moselles, he says : "When properly prepared they are agreeable to the palate and assist digestion ; when new, however, they are liable to communicate their fermentation to the contents of the stomach, and thus seriously disturb digestion." Of clarets and hocks, he says : "They are particularly adapted to table use and as mild stimulants, especially for those predisposed to gout."

ROBERT DUNGLESON, M.D.

Professor Medicine, Jefferson Medical College, Philadelphia.

—"When wine is good and of a proper age it is tonic and nutritive ; when new, flatulent and cathartic, disagreeing with the stomach and bowels. It is perhaps the best permanent stimulus in the catalogue of the Materia Medica." "Beers are nourishing."

THE NATIONAL DISPENSATORY, [PHIL'D. 1879].

By Alfred Stillé, M.D., LL.D., Professor of Medicine, University of Pennsylvania, and John M. Maisch, Phar. D., Professor of Materia Medica in the Philadelphia College of Pharmacy.

—"Experiments and observations made to determine the mode of action of alcohol in the economy have too generally referred to excessive doses rather than to such as are employed in dietetics and medicine. Hence a dispute has arisen whether alcohol raises or depresses the animal temperature, when it is of daily experience that moderate doses augment the heat and excessive doses diminish it. * * * The special effect depends upon the dose and upon the time during which it has been operating."

—"The use of alcohol in every age and by every nation in the world demonstrates that it satisfies a natural instinct, that it literally refreshes the system exhausted by physical and mental labor, and that it not only quickens the appetite for food, and aids in its digestion, but that it spares the digestive organs by limiting the amount of solid food which would otherwise be required. But in accomplishing this salutary end, it does not act as a mere condiment. It is also food, in the sense, at least, that it offers itself in the blood as a substitute for the tissues which would otherwise be destroyed." "Alcohol," says Moleschott, "is the savings bank of the tissues. He who eats little and drinks alcohol in moderation retains as much in his blood and tissues as he who eats more and drinks no alcohol."

—"It is certain that of any amount of alcohol ingested, not in excess of what the system requires, only a minute fraction can be recovered from any of the excretions ; the remainder undergoes metamorphosis."

—"The doctrine that alcohol is food is probably true, not only indirectly by preventing tissue waste, but in the more liberal sense of its favoring the formation of fatty products, especially by increasing the amount of adipose tissue."

—"Habitual excesses, according to the form and quantity of the alcohol consumed, tend to increase fat and diminish muscular tissue, and to render the capillary blood vessels everywhere turgid, giving to the skin, especially to the face, a puffy and purplish look. The digestion becomes impaired, the nervous system unsteady, the gait loses its elasticity, the hands grow tremulous, the senses dull, and the mind torpid ; the sleep is broken, dreamy and unrefreshing. After unusual excesses delirium tremens is apt to occur, and habitual drunkenness sooner or later in many cases induces a state of chronic imbecility with partial paralysis. Fatty heart, cirrhosed liver, and granular kidneys are frequently associated lesions, producing fatal dropsy, especially if distilled liquors have been employed, while gout is the more usual result of intemperance in wine."

—" It would be a gross error to conclude that a given quantity of wine is equivalent to an equal quantity of water containing a like percentage of alcohol. The peculiar effects of a true wine depend upon its ethereal and saline elements, its sugar, tannic acid, etc., which modify materially those which would be produced by alcohol alone."

—" In France drunkenness, as it is known in this country and in the north of Europe, was comparatively rare as long as wine and cider were the only alcoholic beverages iu general use : but the vice has greatly extended of late years, and its increase has been mainly in the cider-drinking departments where alcohol distilled from grain is largely used. In the wine-growing districts where brandy of a less injurious quality is made the use of the latter beverage is comparatively small, and in the departments where wine is not produced the cases of drunkenness requiring legal action are five times greater than in those which consume wines principally. In like manner the proportion of cases of insanity originating in alcoholic drinks is in direct proportion to the use of distilled spirits, and especially of those which are not procured from wine."

—" Wine, including all forms of genuine wine, is essentially a stimulant of the nervous and circulatory systems, but its mode of stimulation is quite unlike that of alcohol. In due proportions it gives to all the faculties increased activity and freedom of action, quickening and brightening the intellect and imagination, warming the feelings, invigorating the digestive powers and diffusing a cordial satisfaction throughout the whole being. Distilled spirits on the other hand tend to benumb the faculties even while stimulating them, so that they seem to be struggling under a brute force."

— " In excessive doses the effects of wine more nearly resemble those produced by distilled spirits, but even these are less extreme and debasing. When indulged in beyond measure its mischievous influence upon the health is less than that of alcohol, for while the latter tends directly to occasion fatty degeneration of the heart, liver and kidneys, the intemperate use of wine is more apt to engender gout, gravel and nervous affections. Even these results are rarely attributable to some of the sorts of wines that are most largely used, as the Rhenish, Bordeaux and most Italian wines."

—" Its (wine's) habitual use is conducive to the health of those who drink it in moderation."

—Referring to the medicinal use of wine in typhoid fever and febrile affections, which assume the typhoid state, the authors say : " It is true of wine as of distilled spirits, that the necessity for its use is often more apparent than real, and that a premature and excessive use of it renders the patients insensible to its operation when it has become essential to his recovery. We have never had such good reason to be plea ed with the results of treatment in typhoid fever as when in a large hospital service the use of alcohol in this disease was almost wholly omitted. In typhus fever enormous quantities have sometimes been administered with the most salutary results."

-- "Wine is often of essential utility in chronic affections which waste the strength by profuse discharges, by pain, or by inducing anæmic conditions."

CARL BINZ, M.D.

Professor of Pharmacology, University of Bonn, in Quain's Dictionary of Medicine.

"There is scarcely any other therapeutic agent the internal action of which varies so much according to the dose given."

"In small quantity and slightly diluted with water alcohol promotes the functional activity of the stomach, the heart and the brain; whilst a like quantity largely diluted exerts but a limited influence upon these organs; if, however, the dose of alcohol be often repeated, it is readily assimilated, and becoming diffused throughout the system, undergoes combustion within the tissues of the body, imparts warmth to them, and yields vital force for the performance of their various functions. Simultaneously with this consumption of alcohol the body of the consumer is often observed to gain in fat, a circumstance due to simple accumulation, the fat furnished by the food remaining unburned in the tissues, because the more combustible alcohol furnishes the warmth required, leaving no necessity for the adipose hydrocarbon to be used for that purpose. A quantity of 100 cubic centimeters of alcohol per diem (about three and a half fluid ounces) equivalent to about one litre of Rhine wine of medium strength, is sufficient to supply between one-third and one-quarter the whole amount of warmth requisite for the human body during the twenty-four hours. The warmth so supplied cannot be measured by a thermometer, however, any more than can that furnished by the internal combustion of other hydrocarbons such as oils or sugars."

--"Under ordinary circumstances and after the consumption of moderate quantities of alcohol only slight traces of it are to be detected in the wine, and none whatever in the breath. Pure alcohol imparts no taint to the exhalations of the body; the ethers of fusil oil on the other hand do so by reason of their being so readily combustible."

--"There can be no doubt but that a healthy organism supplied with sufficient food, is capable of performing all its functions without requiring any especially combustible material for the generation of heat and the development of vital force. But the case assumes a different aspect when in sickness it transpires that while the metamorphosis of tissue goes on with its usual activity, or with increased energy, as happens in many diseases, the stomach refusing to digest ordinary foods fails to supply material to compensate its waste. Here it is then, that a material that can be readily assimilated by the system, and which by its superior combustibility spares the sacrifice of animal tissue, is especially called for, and such a material we have in alcohol. Small, but oft repeated doses of alcohol, largely diluted with water, are generally well tolerated by the weakest stomach, and thus given its absorption and oxidation go on without difficulty on the part of the patient's system."

--"According to the experiments of Dr. Frankland and others the burning of one gramme of alcohol yields sufficient heat to raise the temperature of seven litres of water one degree centigrade, and the burning of one gramme of cod liver oil suffices for nine litres. Now in taking three tablespoonfuls of the oil daily we yield about the same amount of warmth to the body as is given by four tablespoonfuls of absolute alcohol. The oil, however, is digested and oxidized by the organs of the body with difficulty while for the assimilation of the alcohol scarcely any exertion of the working cells is required."

"In this sense alcohol is a *food*, for we must regard as food not only the building material, but all substances which by their combustion in the tissues afford warmth to the animal organism, and by so doing contribute towards the production of vital force and keep up the powers of endurance.

--"In such cases such as where in acute or chronic troubles where the digestive organs refuse to tolerate more substantial nourishment, at least in quantities that would answer the necessities of the case) it is certainly not sufficient to call alcohol merely a stimulant. If alcohol here served merely in the quality of a stimulant its effects would soon pass away, leaving the patient more exhausted than ever, for the human organism is so constituted that it cannot be driven to perform its functions by the application of measures that simply stimulate without supplying some new force to take the place of that put forth by the organs of the body under the impulse of excitement."

--"For *general use* a pure claret, hock or mosel are the preparations mostly to be recommended."

"Amyl alcohol is the agent, to the presence of which, the extremely poisonous action of many drinks upon the nerves and other organs is due. All distilled liquors made from other sources than from grapes contain it to a greater or less extent."

J. PEARSON IRVING, Ph. D.

Lecturer on Forensic Medicine at Charing Cross Hospital. Quain's Dictionary of Medicine.

—"Food and drink can, by their abuse, excite disease, and gluttony is as powerful an excitant as drunkenness, though in temperance outcries this fact is completely lost sight of."

CHAMBERS' ENCYCLOPŒDIA.

(Articles on Diet, Beer, Wine, &c.)

—"From three series of observations made upon himself, Dr. Hammond arrives at the conclusion that "alcohol increases the weight of the body by retarding the metamorphosis of the old and promoting the formation of new tissue, and limiting the consumption of fat." The respiratory and urinary excretions and the faeces were invariably diminished. These effects, occurring when the amount of food was below the quantity required to maintain the weight of the body, were productive of no deleterious effect on the system, but when the food was sufficient to balance the weight from the excretions, and still more, when an excess of aliment over the demands of the system was ingested, the health was disturbed and disease almost induced.

—Hence the laboring man, who can hardly find bread and meat enough to preserve the balance between the formation and decay of his tissues, finds in alcohol an agent, which, if taken in moderation, enables him, without disturbing his health, to dispense with a certain quantity of food, and keep up the weight and strength of his body. On the other hand he uses alcohol when his food is more than sufficient to supply the waste of tissues, and at the same time does not increase the amount of his pl exercise, or drink an additional quantity of water by which the change ti-sue would be accelerated) retards the metamorphosis while an increase amount of nutriment is being assimilated, and thus adds to the plethor condition of the system which excessive food has a tendency to produce. * * * "Tea and coffee are usually believed to have a somewhat similar effect to that which, as we have shown, is produced by alcohol and tobacco. The power of tea in arresting the waste of tissues has, however, recently been called in question."

"Considered dietetically," says Dr Pereira, "beer possesses a threefold property ; it quenches thirst ; it stimulates, cheers, and if taken in sufficient quantity it intoxicates ; and lastly, it nourishes or strengthens. * * From these combined qualities beer proves a refreshing and salubrious drink if taken in moderation; and an agreeable and valuable stimulus and support to those who have to undergo much bodily fatigue."— Article on Food and Drink.]

"Wine is our most valuable restorative when the powers of the body and mind have been overtaxed."— [Ibid.]

"It may e laid down as a general rule that the use of wine, even in moderate quantity, is not necessary for young or adult persons enjoying good average health, breathing fresh country air, and not exposed to overwork, or any other abnormal depressing agency. As, however, life advances, and the circulation becomes languid, wine in moderation becomes an essential, or, at all events, a valuable article of food, and even in early life the physician meets large numbers of townspeople, especially women, engaged in sedentary occupations, who cannot digest the national drink, beer, which is admirably suited to our out-door laboring population, and to persons of higher life who indulge freely in open-air exercise. In such cases the beer is replaced by the more grateful bever age, tea, which, however, when taken too freely, and without sufficient food, often gives rise to a form of distressing dyspepsia, which too often impels the sufferer to seek refuge in spirits. In many such cases cheap wine, which may be purchased under the new tariff at from 1s. 6d. to 2s. a bottle, mixed with an equal quantity of water, will be found an excellent substitute for the beer or tea."— [Article on Wine.]

"The distinctive elements of wine," says Dr. Druitt, ai to be had in abundance in cheap Bordeaux, Burgundy, and other French wines ; in Rhine wine ; in the Hungarian, Austrian and some Greek wines ; and in all with a natural and not injurious quantity of spirit. In prescribing pure wines, viz., light, natural, virgin wine, the practitioner has a perfectly new article of both diet and medicine in his hands."—[Article on Wine.]

"In cases of debility and indigestion, such wine as we are now considering, diluted with cold water may often be freely prescribed with great advantage in place of tea at breakfast as well as luncheon or dinner or supper.—[Article on Wine.]

—"The best of cheap wines are those of Bordeaux ; they are pure, light and exhilarating, moderately strong, seldom containing 10 per cent. of alcohol ; free from sugar and other materials likely to induce gout or headache ; and are admirably adapted, according to Dr. Druitt (who has experimented largely on them) for children with capricious and bad appetites, for literary persons, and for all whose occupations are chiefly carried on indoors, and which tax the brain more than the muscle."—[Article on Wine.]

TREATISE ON FOOD AND DIET.

BY
JONATHAN PEREIRA, M.D., F.R.S., L.S.

" In case of extreme suffering and exhaustion from excessive exertion, and privation of food, the cautious and moderate dietetical use of spirit has, on many occasions, proved invaluable."

" The practice of taking a moderate quantity of mild malt liquor, of sound quality, at dinner, is, in general, not only unobjectionable, but beneficial. It is especially suited to those who lead an active life, and are engaged in laborious pursuits."

REPORT ON ENDOSMOSIS.

BY
PROF. JULIUS VOGEL. M.D., PH. D.

—" If alcohol be diffused with watery fluids through animal membranes, only a small portion of the spirit will pass to the water, while a great deal of the latter will go over to the spirit. This evidently indicates that spirits of wine, when taken into the stomach in a highly concentrated condition, can only reach the blood slowly, and in a diluted state. Thus, the spirit cannot possibly retain its usual action of coagulating albumen, after it has entered the blood of the living organism."—[*Works of the Cavendish Society ; Chem. Reports and Memos.*]

NAVAL HYGIENE.

BY
DR. ALBERT L. GIHON, A.M., M.D., Surgeon U.S.N.
(*Philadelphia, 1871.*)

—" The usefulness of tea, coffee, and alcohol in the form of wine, beer, or whiskey, as food-stimuli or accessory food has been satisfactorily established by Anstie, Lankester and others. * * * * The frightful sequence of intemperate indulgence in alcoholic liquors have resulted in the abolition of the spirit portion of the ration. If the substitution of a pint of beer or half a pint of wine could have been effected there is no doubt of the propriety and benefit of its issue."

STATE MEDICINE.

BY
DE. CHAUMONT, M.D.

—" No single nutritious principle whether nitrogenous or carboniferous, can support life, except for a very short time."
—" Alcohol can be dispensed with by many persons, and might be so by more with advantage. At the same time, in moderation I think it

useful. It ought, however, to be taken considerably diluted, and not in greater quantity habitually than can be destroyed by oxidation in the system. This is on an average from one to two ounces of absolute alcohol in the twenty-four hours, representing one or two pints of beer, or half a pint to a pint of claret, or two to four glasses of sherry, or one or two glasses of brandy. It ought to be taken with meals, generally with the fullest meal, or quite at the end of the day. Occasion may arise for abstinence for one or more days, just as is the case with foods."

POPULAR LECTURES ON FOOD.

BY

DR. LANKESTER.

—" The only remedies that can rationally be employed as medicines are those which act as food on the system."

—" It becomes a matter of serious consideration for those who would put down both drinking alcohol and smoking tobacco, as to whether it might not lead to the objectionable habit— eating opium."

—" We take alcohol as food, just as we administer sal-volatile, camphor, and other drugs as medicines. Medicine and food are more nearly allied to each ot' er t' in most people think, and medicines are constantly administered from a dietetical point of view."

—" Then in regard to the beverages we take, such as tea, coffee, etc.. we find that the greater portion is water, even in regard to beer, taking table beer, which, by the bye, is much the best for ordinary drinking."

PARKES' PRACTICAL HYGIENE.

BY

EDMUND A. PARKES, M.D., F.R.S.,

EMERITUS PROFESSOR CLINICAL MEDICINE, UNIVERSITY COL., LONDON.

Edited by De Chaumont.

--" When alcohol is given to healthy animals in full but not excessive doses the temperature of the body falls. This seems to be shown conclusively by the experiments of Ringer, Richards, Richardson, Binz, Amy, Bouvier and Ruge. In healthy men who have been accustomed to take alcohol in moderate quantities the results are rather contradictory. In a man accustomed to take alcohol Ringer could find no change ; in two men. temperate, but accustomed to take beer and sometimes spirits, I could not detect any raising or lowering of the thermometer either in the axilla or rectum. De Mainzer found no fall of temperature on himself, but a slight fall in another healthy person. Some experiments by Obernier and Fokker are also quite negative. On the other hand Ringer, Binz and Bouvier noticed in some healthy persons a decrease of temperature, and though some of these experiments are evidently rather inaccurate, and though the fall of temperature was inconsiderable, it is difficult to refuse belief that in *some cases there may be a slight* depression of temperature. . . . We may conclude that the effect of moderate doses on temperature in healthy men is extremely slight ; there is no increase, and in many persons no decrease."

—" There is every reason to discourage the use of spirits, and to let beer and wines, with moderate alcoholic power take their place . . . because it is so easy to take them (ardent spirits) undiluted, and thus to increase the chance of damaging the structure and nutrition of the albuminous structures with which they first come into contact."

PHYSIOLOGY FOR PRACTICAL USE.

DR. JAMES HINTON, M.D.

(London, 1874.)

Dr. Hinton is the well known aural surgeon of London, and a well known medical writer. The work which we quote is from a compilation of papers originally published in the *People's Magazine*.

—"Can alcohol fail to be a food when in states of habitual intemperance, in diseased conditions of the body, as in fever, a man lives sometimes for many days together upon spirit and water alone? Surely that thing is rightly called a food upon which life can be maintained."

—"Part of it, (alcohol) we know, is carried off through the alimentary canal as waste," is never therefore taken up into the blood at all ; but the greatest part by far is absorbed into the circulation. Of this a portion is exhaled, a portion also is excreted by the channels above mentioned, (sweat-glands and kidneys), but a portion, and this a large one too, we believe, is broken up into water and carbonic acid, and thus consumed, as Liebig originally surmised, either in the blood, by e action of the red cells, or in the liver, or in the nutrient changes in the interstitial tissues of the body, perhaps in each and all of these places ; or if not thus consumed, it is converted into fat, and thus appropriated to the uses of the economy."

"Two physiological effects, distinct and opposite, have been attributed to alcohol—an early stimulant, a later depressent action. To explain these, several theories have been mooted. All excitation, it was said, means more rapid oxidation, quicker consumption of nutrition, greater wear and tear of tissue ; if the heart's action be quickened by the stimulation of alcohol, this organ must shortly suffer fatigue, the circulation will then become retarded, and the sense of depression be felt. Similarly of the nervous system it was affirmed, exhilaration means what our American cousins call rapid brain-functioning, living too fast for repair ; for the nutritive processes require time and cannot be carried on so quickly as the expenditure of force."

—"There is something true in the theory thus enunciated, and framed no doubt to meet the occasion. Fatigue does follow *over-exertion* of mind or muscle, and is evidence that the processes of repair have not taken place so quickly as those of destruction. But the idea that life and energy are antagonistic things involves fundamental error. Energy, or the manifestation of power, the conversion of force into action, involves no expenditure of life or loss of power. Thinking, or lifting a weight is but a function of tissues provided to issue thoughts or actions. The tissues do not suffer by reason of employment, so long as their nutrition is maintained. Brain and muscle can be very fairly likened to machinery, instinct with life but requiring nourishment, as the engine needs fire and water in order to be put in motion. Both require food in order to perform their functions, but there is this difference between them : The engine may lie by and not suffer, but the nutrition of brain and muscle is requisite that they may live ; and perfect nutrition, the appropriation of what is needful, the removal of what is harmful, cannot be maintained unless the organs themselves be exercised."

—"The employment of his mind or of his hands does not wear out the life of a man, except the work be performed under conditions for which the machinery was never constructed. The human body was made for labor, and parts which are not employed very quickly degenerate ; indeed, we believe more men wear out through indolence, than through strain of overwork."

—"But to revert. The temporary exhilaration conferred by alcohol is termed its stimulant effect ; a sense of well-being is diffused all over the body by the greater fulness of the blood-vessels, and a more equable distribution of animal-heat. Nor is this sense of comfort followed of necessity by any depression, *so long as the alcohol has not been taken in excess;* if however, it accumulate to any extent in the blood, the nervous tissue suffers impaired nutrition and exhibits its fatigue by mal-functioning."

—"As we shall hereafter have occasion to insist. alcohol cannot be called anything else but a food, for it contains elements by which the body can be sustained, but it is not difficult to show that it is a dangerous food, and one that should be very sparingly employed."

—"But it is not merely for its stimulant effects that he (the physician) prescribes, and allows the sparing and careful employment of alcohol-containing liquors—nearly all of them contain other nitrogenous principles as well as alcohol, and are strictly food, with just that amount of stimulant combined with them which makes digestion easy and nutrition more complete. The man who gives up taking beer or wine with his food can maintain himself quite as well without them, but he will eat more bread, more meat, more vegetables, and he will require more time for his meals, *being able to exert himself neither quite so quickly or so vigorously after them.*"

—"One, at least, amongst the many risks attached to fermented liquors, as adjuncts to other food, is their quick assimilation, their very easy appropriation by the body, and the facility with which they staunch the sense of hunger. Thus they are employed often to save time, which a more solid meal would require in consumption ; and the principal meal of the day, which can never be dispensed with if health is to be consulted, is postponed until the evening when the body is fatigued, and requires repose rather than food, and certainly rest before taking it."

—"All foods are stimulants, and all stimulants are foods, and the distinction drawn between them is not merely fanciful but strictly untrue ; furthermore, we might inculcate a doctrine that we are well aware every physiologist would endorse : that *any* food taken in excessive quantity is a poison, that is, acts prejudicially to health, and disturbs and damages the body."

—"Those who inherit an unstable mental equilibrium, irritable, excitable persons, and epileptics, should abstain from alcohol as rigorously as from very poison. Contrariwise, the lymphatic, the over-large, pale, apathetic persons find good wine and beer, not only their best medicine, but a useful adjunct to their dietaries."

PROF. MOLESCHOTT.

(From Circle of the Sciences—Article on Chemistry.)

"Tea and coffee excite the activity of the brain and the nerves. Tea increases the power of digesting the impressions we have received : we become disposed for thoughtful meditation, and in spite of the movements of thought, the attention can be more easily fixed upon a certain object : a sensation of comfort and cheerfulness ensues, and the creative activity of the brain is set in motion ; through the greater collectedness and the more closely confined attention, the thoughts are not so apt to degenerate into desultoriness. Educated persons will assemble to tea for the purpose of investigating a certain subject by a regular conversation, and the higher spirits produced by the tea tend to secure with more facility a successful result."

—"If tea be taken in excess an increased irritability of the nerves takes place, characterized by sleeplessness, by a general feeling of restlessness, with trembling of the limbs ; spasmodic attacks even, in the cardiac region, may arise. The volatile oil of the tea produces heaviness in the head, and in fact, a real tea intoxication, first manifesting itself in dizziness, and finally in stupefaction, takes place. Green tea, which contains much more of the volatile oil than black, produces these obnoxious effects in a far higher degree than the latter."

—"Coffee acts also on the reasoning faculty, but communicates to the imagination a much higher degree of liveliness. Susceptibility to sensuous impressions is intensified by coffee ; the faculty of observation is therefore increased, while that of judgment is sharpened, and the enlivened imagination causes the perceptions more quickly to adopt certain forms : an activity of thoughts and ideas is manifested ; a mobility and ardour of wishes and ideals which are more favorable to the shaping and combination of already premeditated ideas, than to a calm examination of newly originated thoughts."

—"Coffee, taken in excess, causes sleeplessness and a state of excitement similar to intoxication, in which images, thoughts and wishes rapidly succeed each other. A sensation of restlessness and heat ensues, together with anxiety and dizziness, trembling of the limbs and a strong desire to go into the open air. Fresh air is commonly the best means of throwing off this condition, which whilst it continues exercises a really consuming power over man."

—"In Constantinople," says Prof. Moleschott, "the coffee houses alarmed the authorities and were closed." "In the seventeenth century, the coffee houses in London had a similar fate."

—"Spirituous liquors stimulate the imagination in particular. In consequence of this faculty the association of ideas is facilitated and the memory sharpened. The susceptibility of the senses is also rendered more keen, the impressions are quickly and distinctly perceived. The judgment is formed with greater ease, as from the lively idea and the excited memory, the facts from which it is derived are brought nearer together. Hence a frequent and surprising clearness and precision of judgment in matters not requiring a long and close examination. We speak foreign languages with more than usual adroitness. The facilitated movements of thought, the versatility of ideas, are associated with greater ease in the movement of all the voluntary muscles ; the voice becomes fuller and stronger ; the weariness and relaxation following the exertions of the body vanish. Thus arises a feeling of comfort and delight, of increased strength and freshly armed courage ; scaring away ill-humor, sorrow or fear. The affairs of others find a greater sympathy and forbearance."

- "When wine or other spirituous liquors are taken in excess, hallucinations of the senses ensue ; objects appear to the intoxicated person obliterated, blurred or double, &c. At the same time the imagination produces uncertain, variegated, unsteady images, combined without regularity ; the memory refuses its functions in the very act of speaking ; the intoxicated man forgets what he intended to say, and thus the judgment is troubled and confused. Then follow the ebullitions of an unjust passion and a sensibility to opposition, which is so much the oftener irritated as the disturbed action of the brain, over-loaded with alcohol, impairs the correctness of the judgment. Excessive indulgence in wine and all other spirituous liquors causes sleep ; if increased to a complete intoxication the mental functions are disturbed to such a degree that a condition of temporary insanity supervenes ; the senses become blunted, the imagination produces the most various and irregular images which the judgment is unable to examine, distinguish or combine ; all self-possession vanishes, at length even consciousness is lost, the intoxicated man becomes giddy and finally falls into a profound slumber."

—" But before this, a sensation of lassitude and exhaustion arises, the muscles lose their elasticity, the features become flabby, the angles of the mouth widen, &c., and even the respiratory muscles are weakened, the pulse becomes languid and slow."

—"Tea and coffee, though of themselves not difficult of digestion, tend to disturb the digestion of albuminous substances by precipitating them from their dissolved state. Milk, therefore, if mixed with tea or coffee is more difficult of digestion than if taken alone No Italian takes cream with his coffee after dinner. * * Theine is excreted again as urea with surprising rapidity."

—"A good beer is as nutritious as fruit."

—"Alcohol may be considered a savings box of the tissues. He who eats little and drinks a moderate quantity of spirit retains as much in the blood and tissues as a person who eats proportionately more without drinking any beer, wine or spirits."

—" It would be cruelty to deprive the working man who earns his frugal meal by the sweat of his brow of those means by which he is enabled to prolong the nourishment yielded by his scanty food."

—" Fermented liquors, taken in moderation, increase the secretion of the digestive juices and promote the solution of the food. Taken in excess they cause induration of the stomach, which destroys together with the digestive powers, the formation of blood."

—" Sugar is useful to the stomach if it do not by being taken in excess produce too great a quantity of lactic acid."

—"That which renders vinegar so favorite an addition to food is a peculiar acid, consisting of carbon, hydrogen and oxygen, form d from alcohol by the absorption of an additional quantity of oxygen. * * In a majority of cases the use of vinegar is founded on good reasons. * * Beverages containing vinegar have a dissolving effect on the blood and are cooling. * * Vinegar assists digestion. * * In excess it produces dangerous and deeply rooting diseases."

—" Large draughts of water (at meals) would be most injurious with aliments difficult of digestion, like the fats * * but in countries where soup does not constitute a regular part of the meal, drinking water is positively to be recommended."

—" Beer and wine at dinner are hurtful only if taken in excess ; for in the latter case the alcohol coagulates the albuminous substances, not only of the food, but also of the digestive fluid, and thus disturbs digestion. If taken in moderate quantity such beverages are calculated to make the meal hold out longer : for the fact that we are not so soon hungry again after taking a meal with wine, than if we have taken only water with it, is to be accounted for by the slower combustion of the constituents of our body, because the alcohol we have imbibed takes possession of the inhaled oxygen. Hence wine with a meal is extremely useful when a long journey or work in hand renders it impossible to take food again at the usual time ; so much the more so, as such detention from food itself usually causes an acceleration of the metamorphosis of the tissues which beer and wine efficiently obviate."

—"If man eat more than he excretes the tissues become overloaded, which endangers their activity as much as the impoverishment of the blood and the consequent defective nutrition could do ; fat is collected which the oxygen does not consume, and the albuminous substances with the salts assume a fixity which at once enfeebles the intellect and destroys the pleasure of thinking, while it diminishes the strength of the muscles, etc."

—"Vehement, passionate natures become still more ardent from partaking of game, heavy bread, leguminous seeds or any considerable quantity of beer, wine or spirits, coffee or tea. By these more stimulating aliments the circulation is accelerated, etc."

—"Fruit and vegetables with lemonade and similar drinks are more advantageous for irritable constitutions than spirituous and aromatic beverages The latter are more appropriate to persons whose activity of brain is disproportionately great, while their weak digestive organs, their slow formation of blood and nutrition, occasion a disposition to melancholy. Such persons require a stimulating diet."

—"Wines taken in moderate quantity   *   *   accelerate digestion. *   *   They produce a greater uniformity in the functions of the different organs, and thus exercise a beneficial influence upon disposition and character."

—"Where slight irritability is united with flabby muscles, a pale, flaccid, puffy skin, etc., as in phlegmatic persons, a nutritious animal diet is to be combined with strong spices, strong beer and wine ; vegetable aliments, especially roots containing much starch and sugar, must be avoided by all such."

—"A good beer partakes in all the advantages of the alcoholic beverages, and at the same time usefully quenches the thirst by its more abundant amount of water. Hence this beverage is particularly adapted to satisfy the frequent thirst caused by bodily exertion : it is therefore a laudable custom to refresh artizans who have to work hard, in morning and afternoon, with a glass of ale. This beverage by its proportionate amount of albumen, which is equal to that of fruit, supplies even a direct substitute for food."

—"The regular increase in the consumption of alcohol, corresponding to the nature of the climate, cannot but suggest the existence of some valid reason for this popular custom, which has been completly confirmed by scientific investigation. The alcohol which has been taken, is a new source of the development of warmth, by which on the one hand, the food is more slowly consumed ; and on the other, the cushion of fat under the skin which is a bad conductor of heat, and sufficiently protects the system against extreme cold, is kept fron wasting.

—"Inasmuch as man is formed by all the circumstances conjointly, the influence of which upon diet we have endeavored to describe, the rules given must be separately considered. The nature of man is the product, or rather the sum of all those effects of parentage and country, age and sex, position and habits,   *   *   which we have referred to. It is for the reader to determine the choice of food, according to individual influence, bearing in mind that other circumstances to which we have not alluded still remain for consideration. These combinations of circumstances are nearly as numerous as the men themselves, and it must be left to the judgment of the individual to accommodate his diet to his own particular case.

—" Only one point is of the highest value in daily life: intemperance may become the source of the various diseases. This affirmation in its widest sense bears on the use of aliments of every description."

—" The predilection for cooling beverages at this season (summer) has a perfectly rational basis, as the heat frequently produces palpitations, while the less rapid decomposition of the tissues which is peculiar to the warm season, retards the transformation of blood. Cooling and diluting beverages containing vinegar, currant juice, raspberry vinegar and water, operate against this retardation ; but the palpitations are only increased by heating beverages and spices. An allowance of spirituous liquors, is, therefore, doubly injurious in summer, for the alcohol deprives the constituents of the body of inhaled oxygen which is necessary for this composition, and also for the animation of the corporeal functions. For this reason the lighter sorts only of beer and wine ought to be chosen. It is also important to be moderate in the use of heating spices, or to allow them only where a certain excitement of the digestive activity within the limits of health is wished for."

PHYSIOLOGY OF COMMON LIFE.
BY
G. H. LEWES.

—" In compliance with the dictates of physiology, and let me add, in compliance also with the custom of physiologists, we are forced to call alcohol food, and very efficient food, too. If it be not food, then neither is sugar food, nor starch, nor any of those manifold substances employed by man which do not enter into the composition of his tissues. That it produces poisonous effects when concentrated and taken in large doses is perfectly true ; but that similar effects follow when taken in small doses is manifestly false, as proved by daily experience."

—" Every person practically acquainted with the subject knows that concentrated alcohol has, among other effects, that of depriving the mucus membrane of the stomach of all its water, i.e., of hardening it and destroying its powers of secretion ; whereas diluted alcohol does nothing of the kind, but increases the secretion by the stimulus it gives to the circulation."

—" He (Bardleben) found that forty-five grains of common salt introduced at once into the stomach through an opening occasioned a secretion of mucus, followed by vomitings ; whereas five times that amount of salt in solution produced secretion of these effects. The explanation is simple and would be understood by any one who has seen the salt which was sprinkled over a round of beef converted into brine, owing to the attraction exercised by the salt on the water in the beef : this attraction is incalculably small or when the salt is in solution and the salt already saturated."

—" We might multiply examples of the differences which result from the use of concentrated and diluted agents, or from differences in the quantities employed ; as when a certain amount of acid assists digestion, but if increased, arrests it. But the demonstration of such a position is unnecessary, since no well informed physiologist will deny it."

—"Men take their pint of beer, or pint of wine daily for a number of years. This dose daily produces its effect : and if at any time thirst or social seduction makes them drink a quart instead of a pint, they are at once made aware of the excess."

—"We know what a stimulant tea is ; we know that treble the amount of our daily consumption would soon produce paralysis why are we not irresistibly led to this fatal excess.

—"Alcohol replaces a certain amount of ordinary food. Liebig tells us that in temperance families where beer was withheld and money given in compensation, it was soon found that the monthly consumption of bread was so strikingly increased that the beer was twice paid for- once in money and a second time in bread. He also reports the experience of the landlord of the Hotel de Russie, at Frankfort, during the Peace Congress : the members of this congress were mostly teetotallers and regular deficiency was observed every day in certain dishes, especially farinaceous dishes, puddings, etc. So unheard of a deficiency in an establishment where for years the amount of dishes for a given number of persons had so well been known excited the landlord's astonishment. It was found that men made up in pudding what they neglected in wine?"

F. W. PAVY, M.D., F.R.S.

FELLOW OF THE ROYAL COLLEGE OF PHYSICIANS ; PHYSICIAN TO AND LECTURER ON PHYSIOLOGY AT GUY'S HOSPITAL.

From Pavy's Treatise on Food and Dietetics, 2nd Ed. Wood, N. Y., 1881.

—Referring to Dr. Anstie's experiments (1874) which completely sustained the conclusions of Dupre, Thudichum, Schulinus and Baudot against the errors of Lallemand and Dr. Smith, regarding the elimination of alcohol unchanged, Dr. Pavy says :—

—"Evidence is there advanced which shows that only a fractional proportion of the alcohol ingested is eliminated through the various channels of exit from the body. * * An experiment is related in which, after the administration of Bordeaux wine to six persons in sufficient quantity to produce intoxication, not more than one per cent. of the alcohol ingested could be recovered by distillation from the samples of urine."

—"From a review of the evidence as it now stands, it may reasonably be inferred that there is sufficient before us to justify the conclusion that the main portion of the alcohol ingested becomes destroyed within the system, and if this is the case, it may fairly be assumed that the destruction is attended with oxidation and a corresponding liberation of force, unless indeed, it should undergo metamorphosis into a principle to be temporarily retained, but nevertheless, ultimately applied to force production."

—"It appears that about one seventh of the habitual energy—capacity for force production—belonging to nitrogenous matter is carried off by urea and thereby escapes in an unexpended state, where nitrogenous matter is consumed within the body."

—"The weight of evidence appears to be in favor of the affirmative (that alcohol has alimentary value.) A small portion seems undoubtedly to escape from the body unconsumed, but the main part of the alcohol that may be ingested is lost sight of, and presumably from being turned to account in the system."

—"It has been affirmed that the temperature is lowered. Dr. Parkes, however, from his recent thermometric observations, remarks that there is but little change induced in the axilla and rectum of healthy men, but what change occurs is in the direction of increase."

—Referring to the report on the issue of a spirit ration during the Ashantee campaign of 1874, (Churchill's 18, ,) Dr. Pavy says :— "Whilst the general testimony resulted in the condemnation of the employment of *spirits* as a restorative *during* the fatigue of marching : the evidence on the other hand went strongly to show that, issued after the day's fatigue was over, the spirit ration exerted a beneficial reviving effect and afterwards induced an increased feeling of warmth accompanied by the promotion of sleep. Upon these points Corporal Hindley, who had been always a temperate man and never in the habit previously of taking spirits, expressed himself as follows ' Had two rations of rum (a ration equal to 2½ fluid ounces) on the way to the Prah, taken in the evening just before going to bed ; thought it useful ; when there was no issue felt chilly and cold at night : felt warmer when he had taken the rum and slept better ; had no doubt about feeling warmer and sleeping better. On the next day felt no ill effects from the rum.'"

—"Popular belief sanctions the practice which is adopted by many, of swallowing a mouthful of brandy or some other neat spirit after partaking of an indigestible article of food. Now, alcohol consumed in this way, by stimulating the mucous membrane of the stomach and exciting an increased flow of gastric secretions, is calculated in reality to afford assistance to digestion. * * Should it be introduced, however, in larger quantity into the stomach an opposite result is to be looked for. The alcohol now, by virtue of the amount present, will throw down the nitrogenous digestive principle —pepsine —in a solid form and so destroy the energy of the solvent juice. Thus, whilst a small quantity by its stimulant action may assist digestion, a large quantity stops it and accounts for the rejection of food in an undigested state, that is frequently noticed to occur after the too free indulgence in alcoholic liquors at or after a meal."

—"Nothing with such certainty impairs the appetite and the digestive power as the continued use of strong alcoholic liquors. From the stomach the alcohol is absorbed, and with its distribution through the system it interferes with nutrition and leads to a general textural deterioration. Upon certain organs, however, its effects are more manifest than upon others The liver, kidneys and nervous system, for instance, very strikingly suffer, a diseased state being set up, which forms a distinctly recognizable source of death."

—"I have hitherto referred to the action of alcohol *per se*, and in spirits we have little or nothing else, it may be considered, besides this action, to deal with, except, perhaps, in the case of hollands and gin, which possess diuretic properties, due to the flavoring agent (juniper) added. In the primary fermented liquids, however, there are associated ingredients which give rise to the production of modified and additional effects upon the system. The beverages, for instance, which are rich in saccharine and extractive matters, as particularly stout, porter and the heavier ales possess a nourishing and fattening power, which does not belong to a simple alcoholic liquor. Such beverages also are apt to occasion what falls under the denomination of biliousness in those who lead a sedentary mode of life, whilst a lighter and purer alcoholic drink may be found to agree."

—"Beer is a refreshing, exhilarating, nutritive, and when taken to excess, an intoxicating beverage. * * A light beer, well flavored with the hop, is calculated to promote digestion and may be looked upon as constituting one of the most wholesome of the alcoholic class of beverages. It is not all, however, who can drink beer without experiencing inconvenience. In the case of persons of a bilious temperament, &c."

—Cider and perry have, according to analysis by Brande, the following alcoholic strength by volume : -cider, highest average 9·87 per cent : lowest, 5·21. Perry, averaged, four samples, 7·26. They are thus stronger than beer.

—" They (cider and perry, constitute agreeable, wholesome and refreshing, stimulating beverages, when in a perfectly sound condition. Their proneness, however, to undergo acetous fermentation renders it necessary that they should be drank with caution, for in a sour state they are apt to occasion colic and diarrhœa with those who are not in the habit of constantly taking them."

—" The red wines derive their color from the husks of black grapes. These contain the coloring matter in an insoluble form and thus the juice escapes being impregnated with it. Although insoluble in the fresh juice, it, however, undergoes solution when in contact with the fermented juice, the *rationale* being that the presence of alcohol in combination with the acids of the juice gives rise to the production of a liquid possessing a solvent power, which the original aqueous liquid did not. As then alcohol becomes developed in the mixture of husks and fermenting juice, the coloring matter is taken up. The watery juice of the grape, simply impregnated with the acid fails to touch the coloring matter" (of the skins) ; " directly, however, alcohol is present, it becomes taken up. * * * The deeper the color the rougher the flavor of the wine."

—" Except in the case of the teinturier grape (a very rare variety) the juice of grapes is colorless, and hence when wine is made from the juice alone, or with the exclusion of the husk from the fermenting vat, the product is nearly colorless, no matter whether white or black grapes have been employed." For coloring wines "foreign agents, particularly elderberries and logwood, it is asserted, are frequently used.—White wines, it is further alleged, are sometimes dyed."

—" Independently of the amount of saccharine matters in the juice the extent of alcoholic strength is limited by the action of the alcohol generated, for directly a certain quantity is present, a check is put upon the further progress of fermentation and the excess of sugar remains unfermented. Thus, although the juice may have been artificially sweetened by the addition of sugar, or the proportions of sugar increased by the partial dessication of the grape or the evaporation of the juice, only a limited strength can be acquired as the result of fermentation."

—" Natural wine, it may be stated, rarely contains more than 22 per cent by volume of proof spirit." (11 per cent. of absolute alcohol.)

—"About five glasses of the natural and three of the fortified wine contain the equivalent of one glass of brandy."

—" A great deal of unnecessary stress has been attached to the question of the amount of free acid in wine in relation to the production of acidity of the stomach.—Experience shows that it is not acids which particularly favor the acidity of the stomach, but in reality articles containing sugar, especially where the sugar is in an unstable condition.—The presence of a moderate amount of acid does no harm ; on the contrary, it may afford assistance to digestion.--A wine which has acquired sourness from such a source (acetous fermentation) is no longer sound."

--" Whilst wines differ considerably—each should possess a clean, sound and simple taste. It should give an idea of unity in contra-distinction to the mixed tastes belonging to a made up article."

—" A good wine promotes the appetite, exhilarates the spirits and increases the bodily vigor."

—" With the moderately exhilarating and the other properties that the clarets possess they form an exceedingly valuable kind of stimulant, both for the healthy and the sick. There is scarcely any condition in which they are calculated to disagree. They form a most suitable beverage for persons of a gouty or rheumatic disposition, and also for the dyspeptic."

—" It may be said that they " (clarets) "are not prone to turn sour upon the stomach themselves, nor to cause other articles to become sour ; neither do they provoke headache nor derangement in those who are subject to bilious disorders."

—" In character Burgundy is a richer, fuller-bodied or more generous wine than claret.—Therapeutically, it is a valuable agent where poverty of the blood or an ill-nourished state of the system exists. In such cases it is decidedly to be preferred to claret."

—" Champagne has a tendency to allay irritability of the stomach ; unless in a good sound state, however, there is scarcely any wine that is so calculated to upset the stomach." * * * " German wines are of light alcoholic strength,—grateful and refreshing to d.ink, as well as excitant of the appetite." * * Greek wines "constitute natural wines with a high alcoholic strength for their class. The white are clean, fresh and agreeable drinking, whilst the red have fulness and roughness belonging to the bette. kinds and some degree of tartness in the cheaper kinds.—Port, like sherry and marsala, belongs to the fortified class,— pre-eminent as a full, rich strength-giving stimulant—of great service in enfeebled states of the system. For every day use, while suiting many,— far too heavy for others. By dyspeptics, the gouty, &c., should be shunned. Drank in excess—tends to produce plethoric state ; gouty habit may be developed through its influence. Port, some years back, was largely consumed amongst the upper classes,—at the present time its place may be said to be taken by claret. * * Sherry has long held a high position in public estimation as a wholesome and clean drinking wine. Unbrandied sherries are often advertised, but the wine in an unfortified state is only exceptionally in., orted into England and consumed.—Hambro' sherry is a made-up art.cle. Hamburg is not a wine growing but a wine fabricating locality. Much of the cheap sherry sold and a great portion supplied at refreshment rooms and public houses, is derived from this source. It is this which often brings sherry into disrepute by occasioning acidity, headache, and other symptoms of gastric derangement. —Sometimes called Elbe sherry. * * Marsala.—better they should use a good marsala than a bad sherry.—Average alcoholic strength of wines of the fortified class. * * * Madeira is one of the choicest of the fortified wines. * * Mead is a wine prepared from honey and water. Orange wine, currant wine, plum wine, gooseberry wine, &c.,— prepared from fruits ; palm wine, maple wine, parsnip wine, &c.—prepared from other vegetable products ; none will bear comparison for purity and choiceness of flavor with the fermented liquors derived from the grape."

—Brandy—" purity and delicacy of its flavor give it the position it holds and render it suitable for selection in any case where either dietetically or therapeutically, a spirit is required—popular remedy for sickness, diarrhœa, exhaustion, spasms and for correcting indigestion or stimulating the digestion of an indigestible article of food. Burnt brandy is often specially useful in protracted sickness and will be sometimes found to be retained when other articles are rejected. * * Whiskey :—If the flavor be not objected to, whiskey may be used in precisely the same way as brandy. * * Gin possesses diuretic properties, to an extent not enjoyed by the other spirits. Age does not improve it. * * Arrack is a name given to the spirit obtained om a fermented infusion of rice, and also from toddy or palm wine. * * Koumiss, which has lately been extolled as useful in the treatment of consumption, is procured in Tartary from fermented mare's milk, and latterly also, has been made in this country from cow's milk artificially sweetened. * * Absinthe—sweetened spirit flavored with wormwood. Perhaps the most treacherous and pernicious for habitual use of all the liquors of the alcoholic class."

... " Tea is not to be looked upon as constituting an article of nutrition. The quantity of material furnished to the system in the manner it is used is too small to be of significance per se in contributing to the chemical changes which form the source of vital action. If not occupying the position of an article of nutrition, however, its extensive and widely spread employment may be taken as a indication that some kind of benefit is derivable from its use, and it is prob. .y through the nervous system that this is mainly, if not entirely, produced."

.—" In moderate quantities tea exerts a reviving influence when the body is fatigued, but perhaps some of the effect is also attributable to the warmth belonging to the liquid consumed. It disposes to mental cheerfulness and activity, clears the brain, arouses the energies and diminishes the tendency to sleep—to such an extent, it may be, in some sensitive persons as to occasion a painful state of vigilance or watchfulness and sleeplessness."

—"The phenomena produced when tea is consumed in a strong state and to a hurtful extent, shows that it is capable of acting in a powerful manner upon the nervous system. Nervous agitation, muscular tremors, a sense of prostration and palpitation constitute effects that have been witnessed. It appears to act as a sedative on the vascular system. It also possesses direct irritant properties which lead to the production of abdominal pains and nausea. It promotes the action of the skin, and by the astringent matter it contains, diminishes the action of the bowels."

.—" Lehmann was of opinion that it lessened the waste of the body, but Dr. E. Smith asserts that it increases the amount of carbonic acid exhaled, and he thereby speaks of it as promoting rather than checking chemico-vital action. More conclusive evidence, it may be considered, is required in reference to this matter, to show that any decided action either way is exerted."

—" ' Tea' forms an agreeable, refreshing and wholesome beverage."

-- " Its use, particularly green tea, is objectionable in a strong state, in the case of persons who are rendered watchful by it, and in all irritable conditions of the stomach."

—"Coffee forms a favorite and useful beverage. The properties it possesses fully justify the estimation in which it is held. * * * It, however, exerts a more heating and stimulating action than tea, and increases in a decided manner the force and frequency of the pulse. It also differs in being heavier and more oppressive to the stomach. It arouses the mental faculties, &c. Taken in immoderate quantity it may induce feverishness, and various manifestations of disordered nervous action, as tremor, palpitation, anxiety and deranged vision. One of the most valuable properties of coffee is its power of relieving the sensation of hunger and fatigue. * * * In addition to its dietetic value, considerable benefit is often derived from the employment of coffee as a therapeutic agent."

—"The experiments of Lehmann led him to conclude that coffee diminishes the waste of the tissues and causes food to go further, but whether this is true is doubtful. Mr. Squarly could not find that the elimination of urea and chlorides was diminished as might be looked for, if the above view were correct, under the use of large doses of coffee."

PROF. SIR WM. ROBERTS, M.D., F.R.S.

CONSULTING PHYSICIAN TO THE MANCHESTER ROYAL INFIRMARY,
AND PROFESSOR OF MEDICINE IN THE VICTORIA UNIVERSITY.

The following extracts are from Dr. Roberts' address on therapeutics at the Cardiff meeting of the British Medical Association, July, 1885, and are based upon a long series of laboratory observations on the action of tea, coffee, alcoholic drinks, table waters, and other drinks, &c., taken with solid foods, on 1. Salivary digestion, *i.e.*, the action of the saliva as a digestive agent ; 2. peptic digestion, *i.e.*, the action of the fluids secreted by the stomach as digestive agents ; and 3. pancreatic digestion, *i.e.*, the action of the secretion of the pancreas as a digestive agent."

—" In studying the influence of our food-accessories on digestion, it is necessary to distinguish sharply between their action on the chemical processes and their action on glandular and muscular activity. These two actions are quite distinct and generally opposed to each other ; for while all food accessories were found to exercise a more or less retarding influence on the speed of the chemical process, some, if not all of them, exercise a stimulating influence on the glands which secrete the digestive juices, and on the muscular contractions of the stomach."

—" It is also necessary to distinguish between the effects of the food-accessories on salivary digestion and their effects on peptic digestion, inasmuch as wide divergencies were found to exist in this respect. The distilled spirits—brandy, whiskey and gin—were found to have but a trifling retarding effect on the digestive processes, whether salivary or peptic, in the proportions in which they are commonly used dietetically. Their obstructive effects only became apparent when used in quantities which approach intemperance. Taking this in conjunction with the stimulating action which they exercise on the glands which secrete the digestive juices and on the muscular activity of the stomach, their effect in these moderate dietetic proportions must be regarded as distinctly promotive of digestion."

—" In the customary *dietetic* use of wines with meals there is probably a double action ; on the one hand a stimulating action on the secretion of gastric juice, and on the muscular contractions of the stomach ; and on the other hand a retarding effect on the speed of the chemical process. In the case of persons of weak digestion, wines should be taken sparingly, and the quantity so adjusted as to bring out their stimulating action without provoking the retarding effects which follow their more liberal use. * * Effervescent wines, other things being equal, favour the speed of peptic digestion more than still wines."

—" It was found that tea had an inhibitory effect on salivary digestion : even in very minute proportion it completely paralyzed the action of saliva ; on the other hand coffee and cocoa had only a slight effect on salivary digestion. The inhibitory action of tea on saliva was found to be due to the large quantity of tannin contained in the tea-leaf. Some persons have supposed that by infusing tea for a very brief period—two or three minutes—the passage of tannin into the beverage could be avoided. This, however, is a delusion. Tannin is one of the most soluble substances known, it melts like sugar in hot water. * * The effects of tea, coffee and cocoa on peptic digestion were found to be as nearly as possible alike for infusion of equal strength. All three exercised a retarding effect when their proportion in the digesting mixture rose above twenty per cent. These beverages should, therefore, be taken very moderately by persons of weak digestion. * * The strong coffee which it is customary to hand around after dinner must have a powerful retarding effect on gastric digestion ; and although this practice may be salutary to robust eaters, it is not to be recommended to those of feeble peptic power."

—The following extracts from Sir Wm. Roberts are given by Dr. Burney Yeo, in a review in the *Nineteenth Century*, (Feb. 1886) of Dr. Roberts' "lectures on dietetics and dyspepsia" :

—" These generalized food customs of mankind are not to be viewed as random practices adopted to please the palate or gratify our idle and vicious appetite. These customs must be regarded as the outcome of profound instincts, which correspond to important wants of the human economy. They are the fruit of colossal experience accumulated by countless millions of men through successive generations. They have the same weight and significance as other kindred facts of natural history, and are fitted to yield to observation and study, lessons of the highest scientific and practical value."

—"A too rapid digestion and absorption of food may be compared to feeding a fire with straw instead of with slower-burning coal. In the former case it would be necessary to feed often and the process would be wasteful of the fuel ; for the short-lived blaze would carry most of the heat up the chimney. To burn fuel economically and to utilize the heat to the utmost the fire must be damped down so as to ensure slow, as well as complete, combustion. So with human digestion; our highly prepared and highly cooked food requires, in the healthy and vigorous, that the digestive fires should be damped down, in order to ensure the economical use of food. * * * We render food by preparation as capable as possible of being completely exhausted of its nutrient properties : and on the other hand to prevent this nutrient matter from being wastefully hurried through the body, we make use of agents which abate the speed of digestion."

—The following are extracts from Dr. Yeo's summary (*Nineteenth Century*) of Dr. Roberts' conclusions:

—"So far as *salivary* digestion is concerned, these spirits when used in moderation and well diluted, as they usually are when employed dietetically, rather promote than retard this part of the digestive process, and this they do by causing an increased flow of saliva. With regard to peptic digestion the results are still more surprising. It was found that with ten per cent. and under of proof spirit there was no appreciable retardation, and only a slight retardation with twenty per cent. , but with large percentages it was very different, and with fifty per cent. the digestive ferment was almost paralyzed."

—" In the proportions in which these spirits are usually employed dietetically, not only do they not appreciably retard digestion, but these experiments show that they act as pure stimulants to gastric digestion, causing an increased flow of gastric juice and stimulating the muscular contractions of the stomach, and so accelerating the speed of the digestive process in the stomach. For obvious reasons (stated in these lectures) alcoholic drinks, as used dietetically, can never interfere with pancreatic digestion."

—Explaining the retarding action of wines on salivary digestion, he says :
—" This is wholly due to the acid—not the alcohol—they contain, and if the acid be neutralized, as it often is in practice, by mixing the wine with some effervescent alkaline water, this disturbing effect on salivary digestion is completely removed. The influence of acids in retarding or arresting digestion is further of importance in the dietetic use of pickles, vinegar, salads and acid fruits. When acid salads are taken with bread the effect of the acid is to prevent any salivary digestion of the bread, a matter of little moment to a person with a vigorous digestion, but to a feeble dyspeptic, one of some importance."

—" Small quantities of these wines (claret, hock and sherry) do not produce any appreciable retarding effect " (on peptic digestion) " but act as
pure stimulants. These wines, then, may be taken with advantage, even by
persons of feeble digestion, * * in small quantities but not in large."

—-"A moderate quantity of lager beer, when well up, is favorable to
stomach digestion."

—" Beef tea has a powerful retarding effect on peptic digestion. Further
researches appeared to show that the retarding effect of beef tea was due to
the salts of the organic acids contained in it. Beef tea must be looked upon
rather as a stimulant and restorative than as a nutrient beverage., but it is,
nevertheless, very valuable on account of these properties."

" In healthy and strong persons this retarding effect on digestion, observed
to be produced by many of the most commonly consumed food accessories,
answers a distinctly useful end."

I. BURNEY YEO, M.D. F.R.C.P.,

PHYSICIAN TO THE HOSPITAL AND PROFESSOR OF CLINICAL
THERAPEUTICS IN KING'S COLLEGE.

(From " Nineteenth Century" Feb., 1886.)

—" He (man) has departed, in the course of his civilization, very widely
from the monotonous uniformity of diet observed in animals in the wild state.
Not only does he differ from the animals in cooking his food, but he adds to
his food a greater or less number of condiments for the purpose of increasing
its flavor and attractiveness ; but above and beyond this, the complexity of
his food habits is greatly increased by the custom of partaking in considerable quantity of certain stimulants and restoratives, which have become essential to his social comfort, if not to his physical well-being. The chief of
these are tea, coffee, cocoa and the various kinds of alcoholic beverages."

—" It is also a curious fact that many persons with whom tea, under
ordinary circumstances, will agree exceedingly well, will become the subjects
of tea dyspepsia if they drink this beverage at a time when they may be
suffering from mental worry or emotional disturbances."

—" It is a well recognized fact that persons who are prone to nervous ex
citement of the circulation and palpitation of the heart, have these symptoms
greatly aggravated if they persist in the use of tea or coffee as beverages.
The excessive consumption of tea amongst the women of the poorer classes
is the cause of much of the so-called " heart complaints " amongst them ; the
food of these poor women consists largely of starchy substances (bread and
butter chiefly) together with tea, *i. e.* a food accessory which is one of the
greatest of all retarders of the digestion of starchy food."

—" To the strong and vigorous, the slightly retarding effects on digestion,
it " (coffee) " would then " (when a small quantity is taken after a large meal)
" have may be * * not altogether a disadvantage ; but after a spare
meal and in persons of feeble digestive powers, the cup of black coffee
would probably exercise a retarding effect on digestion, which would prove
harmful."

—" The general conclusion to be drawn from these highly interesting and
instructive researches (Sir W. Roberts,) is that most of the food accessories
which in the course of civilization man has added to his diet are, when taken
in moderation, beneficial to him and conduce to his physical welfare and
material happiness, but if taken in excess, they may interfere to a serious
and harmful degree with the processes of digestion and assimilation. It also

is made clear that dietetic habits which may prove agreeable and useful to those who enjoy vigorous health and a strong digestion, need to be greatly modified in the case of those who are feeble and dyspeptic."

FRANCIS E. ANSTIE, M.D., F.R.C.P.,

LATE PHYSICIAN TO WESTMINSTER HOSPITAL AND EDITOR OF THE " PRACTITIONER."

(From " The Uses of Wine in Health and Disease," 1877.)

—" To a medical writer on wines there are several inducements to attack the subject first from the side of the medicinal uses of these drinks ; the strongest reason being that, from the nature of his daily experience, he is most familiar with this aspect of the question. We prefer, however, to start from the view of wine as a beverage of ordinary life ; being persuaded that the subject can only be fairly examined in this way."

—" There is no such clear line between health and disease as is assumed in common speech : the foreshadowings and faint images of disease are to be seen in sundry incidents of the life of those who are conventionally regarded as " healthy," and it is in the study of these natural diseases (if we may use such a phrase) and their relations to the dietary medicines which general custom, independent of medical authority, has prescribed for them, that we are most likely to discover a reasonable basis for the use of these remedies in diseases which involve extensive and obvious departures from the standard of health."

—" The *strong* wines, including port, sherry, madeira, marsala, and all that genus, contain on the average something like 17 per cent. of absolute alcohol, (the strongest ports ranging as high as 23 per cent. or more,) and the light wines, including claret, burgundy, champagne, Rhine and Moselle wines, Hungarian wines, etc., average between 10 and 11 per cent. of absolute alcohol, (the lightest champagnes not containing more than 5 or 6 per cent). Comparing wines with beer, we may note that the poorest sorts of beer contain about 2 per cent. of absolute alcohol ; ordinary table-ale, as drunk in most middle-class households, about 3 per cent. ; ordinary porter between 3 and 4 per cent. ; stout from 5 to 6 per cent. ; while the strongest kinds of malt liquors range through various degrees up to even 10 per cent., and a common strength for good bottled ale or stout is about 7 per cent. of absolute alcohol. On the other hand good brandies and rum average between 45 and 50 per cent."

—" So far as alcoholic strength is concerned it may be said in general terms that half a bottle a day of such wine" (light wines, 10 per cent. alcoholic strength) "for a sedentary and a bottle a day for a vigorous and actively employed adult affords a reasonable and prudent allowance of alcohol ; and this quantity of wine, either clear or with water, will be enough to satisfy the needs of moderate persons for a beverage at lunch and dinner, the only two meals at which alcohol should, as a rule, be taken. * * The same quantity of alcohol represented in beer makes up between two and three pints."

—" We have put this question of the absolute alcoholic allowance for healthy adults in a somewhat crude and abstract form, not undesignedly : for we wish to compel the upper and middle classes and their medical advisers to look the fact of alcoholic consumption honestly in the face."

—" It is therefore much to be desired that people may be educated in the direction of using only one alcoholic drink ; at least for every-day consumption. The choice of this one drink must in each individual case depend upon a number of other considerations besides mere alcoholic strength. What we have practically to consider is the possibility of selecting some alcoholic fluid which shall be weak enough—either when taken neat or with only so much water as will not make it distasteful—to enable us to drink so much of it as will satisfy all need of fluid at lunch and dinner (or dinner and supper with folk of early habits) without producing any of the injurious effects of alcohol. Weak beer would, of course, very well fulfil these requirements : for instance a sound, light, table beer, containing about 3 per cent. of absolute alcohol. But to a large number of persons the quantity of such beer that would satisfy thirst and also prove sufficiently stimulant, would not be readily digestible. Especially to persons of a gouty constitution such a regimen would be most unwholesome ; also to many persons with rheumatic tendency on account of the sugar and dextrine which some light beers contain."

—" It is amongst the class of natural wines, averaging not more than 10 per cent. of absolute alcohol, that we must seek the type of a universal alcoholic beverage for every-day life. * * A bottle a day of either of these wines (9½ per cent. Rhine or 8½ per cent. claret) for an actively employed adult, and a proportionately less quantity for those whose life is more sedentary, would very well represent the allowance of alcohol which may be said to suit best the standard of ordinary health."

—" For the hard-working student, politician, professional man, or busy merchant, there is no better arrangement possible than that of taking, as the regular daily allowance, a bottle of sound, ordinary wire of Bordeaux, and the number of persons with whom such a diet really disagrees is very limited."

—" It is a common idea that the stronger wines are particularly suited to healthy adult life and especially to middle age : but we believe that this is a complete mistake. * * It is otherwise with the two extremes of life—infancy and old age ; in both these periods there are tendencies to a variety of afflictions which scarcely appear to deserve the name of positive disease, but which demand serious modifications of the diet ; these conditions may, we affirm, be far more advantageously treated by the administration of the *stronger* wines than by any other means whatever."

—" It is one of the commonest medical observations that a considerable number of persons can only maintain a good and active state of the digestive powers by means of a very strict limitation of their allowance of sugar and also of the starch-containing foods which undergo conversion into sugar at an early stage of the digestion. Such persons are obliged to be extremely moderate, for instance. in their consumption, not merely of pastry and sweets, but even of bread and potatoes, under penalty of severe dyspepsia if they transgress this rule. To such individuals the *saccharine* wines are very unsuitable and disturbing."

—" Dyspepsia, gout * * * are comparatively uncommon among the European nations who habitually consume the natural acid-tasting light wines in large quantities."

— · One has only to remember the quantities of malic and tartaric acid which everyone swallows during the fruit season, and the quantities of acetic acid which even the most modest consumer of pickles, sauces and salads habitually takes, to perceive how extremely improbable it is' that a wine containing not more than 6 per 1,000 total free acid should (from that cause) disturb the digestion of any moderate drinker of it."

—" It is a singular thing that while the tonic powers of mineral acids, as used in medicine, have obtained universal recognition, the no less remarkable tonic qualities of the vegetable acids, which are essential ingredients of a variety of foods which we consume in ordinary life, should have been so

slightly dwelt upon ; one can only ascribe the fact to the vicious conven-
tional tradition which habitually separates the action of food from that of
medicines."

--" A more profound and important action of this substance (acetic acid)
on the organism is indicated by the empirical observations of numerous
physicians and travellers, that vinegar is an antidote to scurvy. * * It is
probable that malic and tartaric acids, which are the principle representa-
tives of vegetable acid in natural wines, are able to exert a real influence on
secondary assimilative processes."

—" Generally speaking, the tannin element of wine may be said to be neu-
tral as regards its influence on persons in ordinary health."

—" The combination of alkalies (more especially of potash) with vegetable
acids, which every natural wine contains, are of a dietetic value not to be
easily overrated, and in the particular function of warding off scurvy and
some allied diseases of mal-nutrition, they rise to the highest importance."

—" As regards the infancy of delicate children. * * The worthy teetotallers have easily enlisted the sympathies of persons whose experience of
the management of children was limited, when they have declaimed against
the practice of ' rearing drunkards from the cradle,' &c., &c.; and it is, of
course, quite possible to do even so dreadful a thing as this. But the judici-
ous use of wine as a part of the diet even of quite young children, (of course
always under medical sanction) is evidently free from such dangers, and on
the other hand, may do positive good of a very visible kind. The cases
in which it is useful (we are now talking of children not absolutely diseased)
are, (1) those where a tendency to wasting in very marked—*i. e.*, where
children are very apt, without positively seeming ill, to run down suddenly
in flesh with or without simultaneous failure of appetite ; and (2) those where
trifling catarrhal affections are very easily caught and very slowly shaken off.
We are firmly convinced that multitudes of such children have been allowed
to slide into confirmed ill-health, and then into organic disease, who would
have done perfectly well had such symptoms as the above been attended to
by the administration of wine." * * " With children it is much better
to give wine at separate hours, as if it were strictly a medicine."

—" Used under the precautions above given" (as to quantity, &c., &c.) "not
merely is there no danger of corrupting children's tastes, but the services it
renders to health are more important than any medicine with which we are
acquainted ; indeed it is just in the cases where medicines would disorder
the stomach and aggravate the child's *malaise* that wine plays the most remarkable dietetic role."

—" As a dietetic aid in the debility of old age the more potent wines are
even more remarkably useful than in infancy and childhood. More particularly in the condition of sleeplessness, attended often with slow and ineffi-
cient digestion and a tendency to stomach cramps, a generous and potent
wine is often of great value. It is not desirable for such persons to include a
large allowance of fluid in their daily diet, and their alcohol may well be
taken in the more concentrated forms."

Dr. Anstie's advice on when to use and when not to use wine, brandy
and other alcoholic liquors in acute and in chronic diseases, and the safe-
guards against abu e which he steadily keeps in view, cannot be indicated
here by brief extracts with advantage to the reader. In many cases the advantages of these liquors are highly spoken of. In certain conditions of
insanity highly etherized wines " are of inestimable value."

—Regarding certain kinds of dyspepsia where the original cause " is in
truth nothing but nervous depression," Dr. Anstie says, " a fixed moderate
allowance of a generous wine is very helpful. Where we can distinctly make

out from the history that the patient has not exposed himself to the effects of improper food or drink very often such people have been too abstemious in every way) or other ordinary causes of stomach catarrh we may very properly employ a wine of good body and medium alcoholic strength." * * In one form of anæmia, with due discrimination the use of port or other stimulant is commended. In another class of anæmic complaints caused by lack of rest "coupled or not with anxiety of mind" he says, "If this neglect of rest be inevitable, from the press of work, then imperfect as the remedy may be, we believe that alcohol must be allowed and that pretty freely. It must never be forgotten that rest and not alcohol is the true remedy. Nevertheless we are quite certain that it is an error to suppose that alcohol does nothing more than enable such persons to use up their brain tissue faster and thus get work out of themselves for the moment. We cannot doubt that it affords substantial assistance." * * The question of alcohol in phthisis (says Dr. A.) "has engaged our particular attention." "There are two classes in which alcohol appears to play an important part in the arrest of phthisis." The cases, &c., are then described. * * * "It is especially in warding off true tuberculosis from children that the value of wine is conspicuous, and were this more generally recognized we believe that phthisis of children instead of being so fatal a disease as it is, would rarely develop in a fatal form at all : of course supposing that all proper hygienic precautions were adopted and especially a liberal supply of simple and nutritious food. But we repeat here that said in another place : wine should always be given to young children in the form of medicine." * * Certain chronic neuroses of the latter part of life present special aspects in which wine becomes an important consideration, specially a "severe and intractable form of neuralgia rarely or never developed after the age of forty, and one which developed resists remedies with such pertinacity." * * "Still the misery which these neuralgias inflict and the extent to which they shatter the system is deplorable under the best of circumstances, and we need every helpful adjunct we can get. The reflex irritation which the disease sets up is often fatal at once to appetite and sleep, and wine is the true remedy for this part of the mischief."

ENCYCLOPŒDIA BRITTANICA.

(LATEST EDITION.)

(From Article on Diet—Dr. T. King Chambers.)

DIET FOR BODILY LABOUR.

—"The Oxford system of training for the summer boat races may be cited It may be considered a typical regimen for fully developing a young man's corporeal powers to fulfil the demands of an extraordinary exertion, a standard which may be modified according to the circumstances for which training is required :—

—"Rise about 7 a.m. ; a short walk or run, not compulsory. Breakfast at 8.30, of tea as little as possible ; meat —beef or mutton underdone ; bread or dry toast, crust only recommended. Exercise in forenoon, none. Dinner, much the same as for breakfast : Bread, crust only recommended ; vegetables, none (not always adhered to) ; beer, one pint. Exercise—About 5 o'clock start for the river and row twice over the course, the speed increasing with the strength of the crew. Supper at 8.30 or 9 p.m. : meat,—cold ; bread and perhaps a little jelly or water-cress ; beer, one pint. Bed about 10."

—" The Cambridge system differs very slightly."

—" It is notorious that overtraining leads to a condition of system in which the sufferers describe themselves as 'fallen to pieces,' etc. Overtrained persons are also liable to a languor and apparent weakness which is found on examination to depend on the excessive secretion of urea by the kidneys, etc.   *   :   Such are not the results, however, of the training adopted at the universities, by which it would appear that the constitution is strengthened, the intellect sharpened and life lengthened. Dr. John Morgan has collected statistics of the subsequent health of those who have rowed in the University races since 1829, and he finds that whereas at twenty years of age, according to Farr's life table, the average expectation of survival is forty years ; for these oarsmen it is forty-two years.   Moreover, in the cases of death, inquiry into its causes exhibits evidence of good constitutions rather than the contrary, the causes consisting largely of fevers and accidents, to which the vigorous and active are more exposed than the sick.   And it is not at the expense of mind, etc."

—" For reducing corpulency, claret and water at three meals."

—" Small quantities of diluted alcoholic liquids taken with meals slightly increase the activity of the renewal of the nitrogenous tissues, mainly muscle ; that is to say there is a more rapid reconstruction of those parts, as is shown by the augmented formation of urea and the sharpened appetite. Life is fuller and more complete ; old flesh is removed and food appropriated as new flesh somewhat more quickly than when no alcohol is ingested. There appears to be a temporary rise in the digestive powers of the stomach, which is probably the initiative act. The nerve functions are blunted and a lessened excretion of phosphorus exhibits a temporary check on the wear and renewal of the nerve tissues."

—" The effect on a healthy man of taking with a meal such a quantity of fermented liquor as puts him at ease with himself and the world around, without untoward exhilaration, is to arrest the wear of the nervous system, especially that part employed in emotion and sensation. Just as often, then, as the zest for food is raised to its normal standard by a little wine or beer with a meal, the moderate consumer is as much really better as he feels the better for it. When the food is as keenly enjoyed without it the consumption of a stimulant is useless."

DIET FOR MENTAL WORK.

—" When a man has tired himself by intellectual exertion, a moderate quantity of alcoholic stimulant taken with food acts as an anæsthetic, stays the wear of the system which is going on, and allows the nerve force to be turned to the due digestion of the meal. But it must be followed by rest from toil and is in essence a part of the same treatment which includes rest."

—" There is no habit more fatal to a literary man than that of taking stimulants between meals."

—" As to quantity the appetite for solid food is the safest guide. If a better dinner or supper is taken when it is accompanied by a certain amount of fermented liquor, that is the amount most suitable ; if a worse, then an excess is committed, however little be taken."

—" As regards the proper quantity of alcohol that may be used, the two following questions naturally occur : How is a man to know when he has had enough? What are signs of too much?"   *   *   " A more delicate test still is the appreciation of temperature by the skin ; if a draught doesn't chill, if a hot room fails to produce the usual discomfort the wise man knows he has exceeded and must stop at once. In short the safest rule is that when there is a consciousness of any physical effect at all beyond that of satisfaction at the relief of bodily weariness—such a satisfaction as is felt on taking a good meal by a vigorous person—then the limits of moderation have been attained. On ordinary occasions of daily life and ‘for the stomach’s sake’ no more should be taken. *   *   To the practiser of daily temperance, festive occasions are safe and may be beneficial. A man may, from time to time, keep up without harm the above-mentioned sense of satisfaction by good and digestible wine in good company without getting drunk or failure of health, if he makes it a law to himself to stop as soon as he experiences any hurry of ideas or indistinctness of his senses.”

DIET FOR OLD AGE.

—“ In the autumn of life the advantages derived from fermented liquors are more advantageous and the injuries it can inflict less injurious to the body, than in youth. The effect of alcohol is to check the activity of destructive assimilation, to arrest that rapid flux of the substance of the frame which in healthy youth can hardly be excessive, but which in old age exhausts the vital force. Loss of appetite is a frequent and serious symptom in old age, and it usually arises from a deficient formation of gastric juice, which in common with other secretions, diminishes with years. It is best treated physiologically, rather than with drugs.”

DIET IN SICKNESS.

—“ In green sickness, or anæmia, wine is useful at meals on account of the stimulus it gives to the appetite.”

—“ And before quitting the subject of health, as affected by diet, the common sense hint may be given to those who are in good sanitary condition, that they cannot do better than let well alone. The most trustworthy security for future health is present health, and there is some risk of overthrowing nature’s work by over-caring.”

GENERAL.

—“ The social importance of gratifying the palate has certainly never been denied in practice by any of the human race. Feasting has been adopted from the earliest times as the most natural expression of joy and the readiest means of creating joy.”

Signor Cornaro is cited as having, by varying diet and making it consist of such articles as bread, meat, yolks of eggs and soups, lived on 12 oz. per day. “ But he made the solids go so much further by taking 14 oz. of good wine.”

Referring to the dietary regime of one of the European armies the article says : “ The issue of an occasional glass of brandy on holidays makes an agreeable change and benefits digestion, but if wine could be obtained it would be better and not extravagant.”

From Article on Drunkenness.— Dr. W. Balfour.

—“ Apart from the pathological causes, the condition may be actually produced by a multitude of agents, whose use is so widespread throughout the world as inevitably to lead to the belief that their moderate employment must subserve some important object in the economy of nature. Moreover, the physiological action of all these agents gradually shades into each other, all producing, or being capable of producing consecutive paralysis of various parts of the nervous system, but only in doses of a certain amount ; a dose which varies with the agent, the race and the individual. Even the cup so often said ‘ to cheer, but not inebriate ’ cannot be regarded as altogether free from the last named effect. Tea sots are well known to be affected with palpitation and irregularity of the heart, as well as with more or less sleeplessness, mental irritability and muscular tremors, which in some, culminate in paralysis, while positive intoxication has been known to be the result of the excessive use of strong tea." (See 3rd Annual Report of the Massachusetts Board of Health, page :29.)

—" It is well to remember that there is not a shadow of a proof that the moderate use of any of these agents as a stimulant has any definite tendency to lead to its abuse ; it is otherwise with their employment as narcotics."

—" It is interesting to know that a late judge, who lived to nearly ninety years of age, believed he had prolonged his life, and added greatly to his comfort by the moderate use of ether, which he was led to employ because neither wine nor tobacco agreed with him : while the immediate use of the same agent has, particularly of late and in the north of Ireland, given rise to a most deleterious form of drunkenness."

—" However degrading, demoralizing and pauperizing the vice of drunkenness may be, it is important to remember in all our thoughts concerning it, that it is the outcome of a craving innate in human nature, whether civilized or savage, and that there has been no period in the world's history and no nation on its surface in which one or other, and often several simultaneously, of the many natural or artificial nervine stimulants have not been employed, and well has it been for those who have used them moderately."

—" It may be, as a recent speaker has said, that 'a national love for strong drink is a characteristic of the nobler and more energetic populations of the world,' it may be, as he goes on to say, that it 'accompanies public and private enterprise ; constancy or purpose, liberality of thought and aptitude for war ; it,' as he further adds, 'exhibits itself prominently in strong and nervous constitutions and assumes in very many instances the character of a curative of itself.' In other words, in certain constitutions the moderate use of stimulants excites to action rather than to a sensual keyf ; and the pleasurable stimulus of action renders such individuals less likely to fall into degrading habits of excess."

—" The desire for stimulants is one of the strongest instincts of human nature. It cannot be annihilated, but may be regulated by reason, conscience and by law when it encroaches on the rights of others."

DR. DUPRE'S CONCLUSIONS.

(1872.)

The experiments of Lallemand, Perrin and Duroy published in an elaborate memoir in 1860, were interpreted as showing that alcohol was eliminated unchanged. The conclusions and the errors in experimenting which led to them were subsequently pointed out, and it was shown how unfounded the conclusions were by a series of experiments by Dr. Dupré and others. Some of Dr. Dupré's conclusions are here summarized :—

—" The amount of alcohol eliminated per day does not increase with the continuance of the alcohol diet, and therefore all the alcohol consumed daily must of necessity be disposed of daily, and as it certainly is not eliminated within that time it must be destroyed in the system."

—" The elimination of alcohol following the ingestion of a dose or doses of alcohol, ceases in from nine to twenty-four hours after the last dose has been taken."

—" The amount of alcohol eliminated in both breath and urine is a minute fraction only of the amount of alcohol taken.."

J. MILNER FOTHERGILL, M.D , EDIN

PHYSICIAN TO THE CITY OF LONDON HOSPITAL FOR DISEASES
OF THE CHEST

(From "Manual of Dietetics."—1886.)

—"As to the food value of vinegar it is not easy to find anything positive. No doubt it is changed and oxidized in the body ; and it is agreeable to the palate as an addition to many articles of food ; but beyond that it has a further utility. 'It seems not improbable that as vinegar powerfully excites the secretion of the salivary glands, it exerts a similar influence upon the stomach, augmenting the flow of gastric juice and thereby increasing the digestive powers of the organs.' (Stillé). Of its evil effects when taken to excess there is no question."

—"Condiments are agreeable to the palate, and in moderation, good for the digestive organs."

—"Whether his (Sir W. Robert's) conclusions will be entirely confirmed by other observations or not, we may feel pretty confident about the main facts. The well known effects of ' high tea ', *i. e.*, tea with a solid meal, seem to find its explanation in the effects of tea retarding digestion and especially salivary digestion. If the digestion of starch be thwarted and a quantity of undigested starch be lying on the stomach, no wonder if pains and still more flatulence be experienced."

—"The founder of high tea was certainly not a benefactor of his species ; nor indeed has the introduction of tea from China—a static country—into England, a country of distinct advance, and from thence wherever the Anglo-Saxon has spread himself, been an unalloyed good. Tea and damper (a cake of dough without leaven or yeast) may be a meal of simple character and easily prepared, but nothing can be said for it in a dietetic point of view. No wonder that indigestion is as common among the stock riders of Australia as among those of Mexico. Tea may be drunk as a refreshing stimulant beverage, but it is not desirable as a food accessory, especially where the food is largely farinaceous."

—"That wines and malt liquors have an action upon the stomach which is pleasant and grateful is a fact most people know for themselves. That they stimulate the gastric flow and increase the muscular activity of the stomach is probably the underlying cause of the sense of well-being which accompanies their use."

—"It (beef tea) is not a food ; it is a stimulant. Grateful and acceptable alike to the palate and stomach, possessing stimulating qualities, beef tea has its value. But, all the same, as regards its food value it is but a jackass in a lion's skin."

—"The 'tea drinker's heart' is a well recognized malady. People, especially of the female sex, who drink largely of tea and still more when they do not eat a sufficient quantity of food, suffer from many nervous troubles as well as palpitation and neuralgia and present all the phenomena of nerve exhaustion. The nurse often drinks tea to sustain her when there is no appetite for food until utter exhaustion is the untoward result and a state bordering on the delirium tremens of alcoholic excess is revealed. The tea drunkard is a well-known patient at all hospitals ; and is not unknown in private practice. * * 'The cup which cheers but does not inebriate' is not without a toxic influence when taken to excess."

—"Total abstainers are ranked in battle array on the matter with other observers. These last, and their names are both numerous and weighty, hold that alcohol is largely burnt in the body by oxidation and is therefore a 'fuel food.' Personally, after very considerable attention to the subject, I must say that I am among those who hold that the chief portion of the alcohol ingested undergoes consumption in the body."

—"As to the food value of alcohol in health it is utterly subordinate to its action on the nervous system. But in ill-health, and especially in acute disease, the question of its food value as well as its value as a stimulant may well engage our attention."

—"Alcohol requires no digestion."

—"The well-fed navvy who prefers to do the last hour's work of his laborious day on a pint of good sound ale, has a hard, solid argument on his side."

—"There are two matters connected with alcohol well worth bearing in mind.—(1) Never have alcohol on the brain when it has work to do. (2) A little alcohol betwixt a man and a past trouble is permissible, but it is not well to put a little alcohol in front of a coming trouble."

—"All matters which are oxidizable in the body are 'foods.'—*Hermann.*

T. LAUDER BRUNTON, M.D., D.Sc., F.R.C.P., F.R.S.

From the "Lancet's" report of the Lettsomian Lectures, delivered before the Medical Society of London in 1885.

—"Ludwig and Schmidt-Mulheim have shown that peptones when ingested into the general circulation are poisons, producing loss of coagulability in the blood and great depression of the circulation, so that even the products of healthy digestion might prove fatal if they passed rapidly into the general circulation."

—"Alcohol is one of the most powerful stimulants both to secretion and the circulation. Hence the value of the pre-prandial sherry and bitters, of the sherry with soup, of the champagne with its effervescing carbonic acid and of the small glass of liqueur or brandy and the like."

—"Provided that moderation has been exercised, no harm will result even from a heavy meal. Moderation is a relative term. The symptoms of gastric indigestion—loaded tongue, loss of appetite, tendency to nausea and perhaps even vomiting—on the next day, result if the food has been excessive in quantity and more especially if alcoholic stimulants have been taken in excess. Dr. Beaumont, in the case of Alexis St. Martin, said that there were 'several red spots and patches abraded of the mucous coat, tender and irritable, when the above symptoms were present.' The gastric juice then seems to have an alkaline reaction and very little or no digestive power ; hence undigested lumps may pass on and cause diarrhœa, or the inflammation may extend to the intestines ; then dulness and languor with irritability supervene, &c."

—"Man is a low-pressure engine and works almost all his organs considerably under their full power."

—"Cookery may not only be a powerful moral agent in regard to individuals, but may be of great service in regenerating a nation. Schools of cookery for the wives of workingmen in this country will do more to abolish drunkenness than any number of teetotal associations."

—"Walking or other exercise after a long day's work and prior to dinner, by exhausting the nervous system, was a potent cause of indigestion. Effects somewhat similar may be produced by disturbing mental emotions or bodily conditions."

—"In regard to the use of alcohol in dyspepsia, Dr. Brunton thought St. Paul's advice to Timothy was very good : 'Drink no longer water, but use a little wine for thy stomach's sake and thine often infirmities.'"

(From text of Lettsomian Lectures, Published 1886.)

—"A succession of heavy dinners is no doubt injurious; but when the organism is exhausted, a good dinner with abundance of wine is sometimes of the greatest possible use."

—"The alkaloids formed either by normal digestion or by abnormal putrefactive process in the intestine, might readily pass to the heart, nervous system and kidneys, and cause dangerous or fatal consequences."

—"A glass of cold water, slowly sipped, has more effect upon the pulse than a glass of brandy swallowed at a draught. It lasts while the sipping is continued, if the sips be taken at short intervals, but it passes away after the sipping ceases. While its effects upon the pulse is thus greater for the time than that of alcohol, it is much less permanent."

—"When pushed beyond a certain point the appetite rebels, and the 'full soul loatheth the honeycomb;' but before this point is reached a good deal more than enough may have been eaten; and if the same process be repeated every day serious mischief will undoubtedly result; and the more accommodating the appetite is the more serious will the mischief be. Many a man has been saved by a weak stomach. * * Where the stomach and intestines are more accommodating and continue to digest all that is put into them, the burden of the work is shifted elsewhere, and either the liver fails to reconstruct the new material with which it is deluged or the tissues are poisoned and the overworked kidneys become degenerated."

—"Heat is a powerful stimulant to the heart, and a cup of hot tea is therefore much more stimulating and refreshing than a cold one. * * The practice of sipping the tea almost boiling hot is, however, apt to bring on a condition of gastric catarrh."

—"Another cause of imperfect digestion is fatigue. * * But where we are tired the case is very different; a little roughness in the road will cause us to stumble and an unexpected stone may give us a sudden fall. The nervous system no longer co-ordinates the movements of the various parts of the body so that they no longer work together for a common end. The same thing occurs with the various parts of the intestinal canal. * * The acts of chewing and swallowing appear to act as stimulants to the circulation and nervous system, and thus to ensure the proper co-ordination between the functions of the mouth, stomach, intestines and liver. But, if the nervous system be exhausted by previous fatigue, or debilitated by illness, the requisite co-ordination may not take place, and indigestion or biliousness may be the result. Effects, somewhat similar to those of fatigue, may be produced by depressing and disturbing mental emotions and bodily conditions."

—"Some would deny that it (alcohol) has any place at all, and assert that it is injurious at all times and in all places. But such assertions are valueless; they contradict the common experience of mankind and defeat their own end by their extravagance. It is no use to deny the existence of facts, for they will continue to be facts, whether we allow them or not. What we have to do is to open our eyes to their existence and regulate our conduct accordingly."

—"It is impossible to lay down a rule for the quantity necessary, for this will vary, not only with every individual but with the individual at different times. The stimulant which is most generally useful is claret."

—"I do not think it a sin to use alcohol, in moderation, as a luxury, provided always that it be used in moderation, not only for the individual, but for the individual at the particular time at which it is taken, for what is moderation at one time would be excess at another."

(From Paper on the Physiological action of Alcohol.)

—" The latter "(heat, &c.,)"produce permanent coagulation while the coagulum formed by alcoho: readily dissolves again in water, or in the liqui: of the body * * counteracted by the blood which dissolves the albumt . as fast as it is coagulated, so that we do not see any opacity of the mucous membrane o' the mouth, unless alcohol has been acting upon it for a good while. * * The coagulation of albuminous fluids by alcohol seems due in the first instance to the simple abstraction of water, and when this is added again they re-dissolve. If the alcohol acts for a long time upon them, however, their constitution seems to undergo a change and they become insoluble in water."

—" Most substances which possess the power of coagulating albumen * * act as astringents when taken internally. * * Alcohol is no exception to the rule and we all know that a person suffering from an attack of diarrhœa usually flies to the brandy bottle for relief before he thinks of consulting a medical man."

—"When the stomach is empty its mucous membrane as seen through a gastric fistula, is pale and only covered with a little mucous. If a little alcohol is now introduced the blood-vessels of the mucous membrane dilate and it becomes of a rosy red color, its glands begin to secrete copiously, beads of gastric juice stand upon the surface, become larger and larger until they can no longer preserve their form, when they coalesce and run down together in a little stream."

—"How is it that we take a glass of spirits with our lobster to digest it ? Is not this adding fuel to fire and increasing the irritating effects of the lobster on the stomach by that of alcohol? By no means—the fibers of lobster are probably in themselves no more irritating than fibers of beef, but only less soluble in gastric juice, so that they retain their form and hardness instead of being reduced to a pulp, and by thus exerting for a longer time a mechanical irritating action upon the stomach, they produce nausea and indigestion, not immediately after they have been swallowed, but in the course of some hours. If, however, an increased secretion of gastric juice be produced by means of a glass of spirits swallowed at the same time with the lobster, we may expect that digestion will take place more rapidly, the fibers will be dissolved and the prolonged irritation of the stomach being avoided no nausea will ensue."

—" In patients convalescent from an acute illness, or weak, delicate anaemic persons, the food does not sufficiently stimulate the weakened stomach, the secretion of gastric juice is small and the meal lies for a long time like a weight at the epigastrium. The same is the case with the merchant, the lawyer or the doctor, who comes home from his counting house, his office or his rounds and sinks exhausted in his easy chair, weary and worn out by a long day's work. In such cases the diminished sensibility of the stomach must be compensated by an extra stimulus and the glass of sherry which to a healthy person not exhausted by over fatigue would be superfluous, will in them restore the normal equilibrium and quicken the otherwise slow and imperfect digestion."

—" Alcohol taken into the stomach increases the movements of the organ as well as its secretion, and by mixing the contents more thoroughly with the gastric juice accelerates digestion."

—Of the reflex action on the heart and vessels of *large* doses of alcohol introduced into the stomach : " The irritation produced by it is conveyed by the different nerves to the medulla oblongata and thence by the vagus to the heart which it either slows or stops entirely. The mode of action on the intestines is not certain. It may simply arrest the normal action of the vaso-motor centre upon the intestinal vessels, or may be conducted down to them by vaso-inhibitory nerves. In either case it will cause them to dilate." Of the reflex action of *moderate* doses :—"The irritation is conveyed to the medulla, but instead of calling into action the vagus and vaso-inhibitory nerves it excites the accelerating nerves of the heart and probably the vaso-motor nerves of the intestines, thus increasing instead of diminishing the circulation in the body generally. This difference in the   flex action of large and small doses of alcohol upon the heart an   sels corresponds to the different action already noticed, of slight   great irritation of the stomach, mechanical or otherwise, the slight stimulation increasing and the great diminishing or arresting the circulation and secretion."

—"It is the merit of Baudot, Anstie and others who have worked at this subject, to have shown that alcohol is oxidized and is thus to be reckoned as a food and not merely as a drug. But still more satisfactory evidence of its claim to the title of food is afforded by the fact that it will keep up the weight of the body and prolong life when the supply of other food is insufficient or entirely wanting."

—"From a survey of all the evidence on the subject I think we may conclude that in moderate doses alcohol undergoes combustion in the body and will supply energy, yield warmth and tend to sustain life in the same way that sugar would do, and is therefore to be reckoned as a food."

—"In order to prevent any complication, from the reflex action of which we have already spoken, let us suppose that instead of pure brandy, diluted spirits or some light wine has been taken, which will have little or no irritating effect upon the gastric membrane. One of the best possible opportunities of studying the earlier and slighter effects of alcohol is afforded by a public dinner. If we look at our hands and those of our neighbors before going in, especially if the ante-room is somewhat cold, we may find them somewhat pinched looking ; the colour somewhat dusky and distributed in patches instead of being uniform ; the veins very thin, almost like threads. They are of a somewhat blue colour and on emptying them by pressure they fill very slowly, showing that the circulation is languid. After a few glasses of wine, however, their appearance begins to change. The hands now assume a uniform rosy tint, showing that the capillaries are now dilated and filled with bright arterial, instead of dark venous blood ; the veins swell up, become prominent, of a light blue colour almost like arteries ; and when emptied by pressure fill rapidly, showing that the circulation has become very quick, and that they, like the capillaries, are now filled with blood which is tolerably bright, if not quite arterial, instead of the dark blood they previously contained. The hands entirely lose their shrunken look, &c. This dilation of the vessels, so readily seen in the hands, is not confined to them but occurs generally throughout the body. The warm blood pouring from the interior of the chest and abdomen, over the surface, imparts to it a pleasing glow and a most agreeable feeling of comfort pervades the whole frame. The face shares the general flush, &c. * * The muscles acquire new strength ; the work which previously fatigued them is done with ease ; the mental faculties become much more acute and new ones, previously unsuspected, may even appear, &c. * * Provided the liquor has been good, or in other words, provided the alcohol employed has been free from all injurious admixtures, all these effects, I believe, may be produced and may pass away without any bad effects."

—"The slight diminution in the oxidizing power of blood, which alcohol occasions, is many times over compensated by the amplitude of its current."

—"Exercise, like alcohol, both dilates the vessels and increases the action of the heart.

—"While alcohol is thus injurious during prolonged exposure to cold, the case is very different after the exposure is over, and its administration may then be very beneficial. * * The cutaneous vessels, so long contracted by the cold, will not relax all at once, and the deeper tissues gain heat very slowly, just as they lose it very slowly, by mere conduction through 'he skin. * * If a little spirits be now taken * * the cutaneous vessels dilated allow the blood to circulate through them, and become warmed by the fire it returns warm to the internal organs. * * At the same time the dilation of the cutaneous vessels opens new channels to the blood which has been pent up in the interior of the body, and thus lessens any tendency to congestion or inflammation of internal organs, so that a glass of hot brandy and water, at the proper time, may possibly prevent bronchitis or pleurisy."

—"Although alcohol during continued exposure is generally injurious, yet in some instances when pain or cramp in the internal organs seems to indicate more risk from the engorgement than from diminution of the general temperature of the body, it may be beneficial even while the exposure continues."

PROF. SCHÜTZENBERGER.

DIRECTOR AT THE CHEMICAL LABORATORY AT THE SORBONNE.

—" The examination of its " (yeast's) " biological functions, studied more particularly in their chemical aspect, shows us clearly that this elementary form of life does not differ in essentials from other elementary cells unprovided with chlorophyll, whether isolated or in groups, and belonging to the more complex organs. It breathes, transforms and modifies its proximate principles in a continuous manner, and cer'ainly in the same way as other cells ; like them, it can be multiplied by buds and spores. The only important and decidedly distinctive character which seems to render it a form of life, absolutely apart from other forms in creation, was removed from it by M. Lechartier and M. Bellamy, when these chemists succeeded in establishing that the cells of fruits, seeds and leaves, and even animal cells are capable of changing sugar into alcohol and carbon dioxide."

—"Admitting, with M. Pasteur, and all those who have oreceded and followed him in this inquiry, that alcoholic fermentation, and the other phenomena of the same order, * * are palpable manifestations of certain physiological functions of ferments, or organisms of an inferior order, we may ask, whether the power of resolving glucose into alcohol and carbon dioxide, or of changing it into lactic acid, and that again into a mixture of hydrogen, carbon dioxide, and butyric acid, belongs, for each special fermentation, only to a single organism, to a single ferment, or at least to species very nearly allied, as we have seen in the species of the genus *Saccharomyces ;* or whether these reactions are the result of cell life in general, when organic cells are placed under special conditions. On this hypothesis, ferments would have no advantage over other living cells, than that of showing these manifestations in a more energetic and intense degree."

—"There is no doubt that this thought must have presented itself involuntarily, as we may say, to the minds of men of science accustomed to reflect on these delicate questions of the reactions in living organisms, but M. Pasteur has the honor of first clearly expressing it, and supporting it by positive experiments. (Pasteur, Comp. Rend. de l'Acad. des Sciences, Vol. 75, p. 784.) These experiments were undertaken in order to prove that the alcoholic fermentation of sugar may be excited by other organisms than the cells of *Saccharomyces*" (yeast) "and notably by the elementary cells of larger plants, such as are found in fruits, leaves, &c. The researches of M. Lechartier and M. Bellamy, on the alcoholic fermentation of fruits * * have been directed to the same end and lead, as we shall see, to this important consequence, that the elementary organs of plants in general are endowed, though in a less degree than the cells of yeast, with the property of exciting alcoholic fermentation."

—" The same investigator " (M. Fremy) "observed the production of alcohol and carbon dioxide in the interior of fruits, such as pears and cherries."

—" This observer," (M. Lechartier) " placed fruits (pears, apples, lemons, cherries, chestnuts, potatoes, grains of wheat, linseed, currants,) in test jars, connected with smaller test tubes, placed over a mercurial trough. Under these conditions, the whole of the oxygen of the confined air in which the fruits are placed is absorbed. This absorption is accompanied and followed by a considerable production of carbon-dioxide gas. The disengagement of carbon-dioxide is generally divided into two distinct periods ; in the first, after the absorption of the oxygen gas of the air which remained in the test glasses, the disengagement of carbon-dioxide proceeds, at first, in a regular and uniform manner, then it slackens, and stops for a certain time, to be afterwards renewed with an increasing rapidity, greater than that observed during the first period. This continues for several months. At this time, if we have taken the precaution to experiment on fruits isolated from each other and kept from contact with the sides of the vessel containing them ; if, besides, we have been careful to prevent any deposit of liquid on the surface of the fruit, we may ascertain the production of notable quantities of alcohol, easily separated by distillation, after the fruit that has been experimented on has been crushed into pulp ; and what is more, a careful microscopical examination of the parenchyma reveals no trace of alcoholic ferment."

—" Thus, on November 12, two pears, one weighing 157 grammes, and the other 125, were suspended separately, each in a glass jar, well corked, and provided with a tube by which gas might escape. Calcium chloride had been previously placed at the bottom of the jars, to maintain around the fruits an atmosphere unsaturated with the vapour of water. The test jars were opened on July 19 : 1,762 cubic centimetres of gas and 2·62 grammes of alcohol were collected. The pears had preserved their colour, their skin was wrinkled, but not damp. Their consistence and smell were like those of mellow pears. They had lost together 134 grammes of water, yet they still contained it in the proportion of 69 per cent. of their weight. Microscopical observations, made at different distances from the centre, could discover no alcoholic ferment."

—" It appears, then, as these writers say, that at the moment when fruits, seeds, and leaves are detached from the plant which bears them, life is not extinct in the cells of which they are composed. This life goes on, sheltered from the air, consuming sugar and producing alcohol and carbon-dioxide. The moment when carbon-dioxide ceases to be formed, is that at which all the vitality of their cells is destroyed. Fruits, seeds and leaves may *then* remain for an indefinite period in an inert state if no organic ferment is developed in the interior.'

—" 'Guided by all these facts, I have been gradually led to look upon fermentation as a necessary consequence of the manifestation of life when that life takes place without the direct combustion due to free oxygen.' "— *Pasteur.*

—" They prove irrefragably that vegetable cells can produce alcohol and carbon-dioxide at the expense of sugar, and can act like the yeast of beer, but much less energetically ; but nothing in the description given of their experiments by M. Pasteur and M. Lechartier can induce the reader to admit that this production of alcohol, this cellular fermentation, only commences from the very moment when the cell is either partially or entirely removed from contact with oxygen."

MISCELLANEOUS EXTRACTS.

—" In no part of Germany do the apothecaries' establishments bring so low a price as in the rich cities on the Rhine, for there wine is the universal medicine for the healthy as well as the sick ; it is considered as milk for the aged."—*Baron Liebig.*

—"It is true that thousands have lived without tea or coffee, and daily experience teaches us that under certain circumstances they can be dispensed with without disadvantage to the merely animal functions. But it is an error, certainly, to conclude from this that they can be altogether dispensed with in reference to their effects ; and it is a question whether, if we had not tea or coffee, the popular instinct would not seek for and discover the means of replacing them. Science, which accuses us of so much in this respect, will have in the first place to ascertain whether it depends upon sensual or sinful inclinations merely that every people of the globe has appropriated some such means of acting on their nervous life—from the shore of the Pacific, where the Indian retires from life for a few days in order to enjoy the bliss of intoxication with koka, to the Arctic regions where Kamschatdales and Koriakes prepare an intoxicating beverage from a poisonous mushroom."—*Baron Liebig.*

—"As the action of alcohol, in dietetic doses, on the elimination of nitrogen and on the bodily temperature is so utterly negative, it seems reasonable to doubt if alcohol can have the depressing effect on the excretion of pulmonary carbon, which is commonly attributed to it."—*Dr. Parkes on his experiments.*

—"There can be no doubt that many have found by personal experience that they could undergo more bodily and mental fatigue when drinking no draught but the crystal lymph, than they could do when using wine, beer, spirits, and whatever else, in the way of beverage, was deemed generous and nourishing. On the other hand no one can look at the positions ir which man is placed in civilized countries, such as ours, at the habits which have been engendered from infancy amongst the population, at the great variety of constitutions and states of health amongst individuals, and at the wear and tear of mind and body, so universal amid the struggles and competition characteristic of go-ahead communities, without seeing that stimulants must be resorted to by many, in spite of everything that can be said against them. And amid this endless variety of circumstances and contingencies in which we are placed, and the sudden prostrations of mind or body to which we are liable, to say that tea, coffee, wine, beer and spirits are only injurious and never of service, would be to contradict an experience almost universal from the Deluge to our own time. We would say, then, that stimulants being used * * the weaker only should be so habitually, and always in moderation. The stronger should be reserved for emergencies. The habitual use of ardent spirits is altogether to be deprecated, even by non-teetotallers. When used it should be in some emergency, when they may serve a temporary purpose, but there the use of them should cease. There is, and can be, no valid objection against the moderate use of pure or light wines, or yet against wholesome beers. They do not tempt to excess and they refresh and exhilarate. Brandied wines are open to the same objection as ardent spirits in a rather less degree. They should be used but sparingly.—The injurious results flowing from the use of narcotic and stimulating beverages are singularly aggravated by the adulterations practised in regard to them.—Tea and coffee, those '*slow poisons*' of which so much was written *pro* and *con*, when the present century was first looming in the distance, by the 'Sir Sapients' of that day.—We are taught that 'every creature is good, and nothing to be refused if it be received with thanksgiving.' Philosophy confirms this teaching."—*Baird's Management of Health.*

—" Much discussion has been expended on its elimination and combustion, and some years ago the hopes of the 'temperance' agitators were much raised by the important results of experiments put forward by two French observers which seemed to prove that all the ingested alcohol is given out unchanged in the urinary and other secretions. Anstie and Dupré, however, showed the fallacy of this and pointed out that a substance is eliminated in the urine of the most rigid abstainers which cannot be distinguished from alcohol, and it has since been asserted that this may actually be alcohol derived from converted liver sugar."—*Dr. Farquharson, M. D., Edin. ; F.R.C.P., London.*

—" On the other hand it is necessary to remember how often the whole question must practically be approached from a totally different aspect ; how often alcohol constitutes not the single feather which distracts the sleepy savage, but the bed of down which restores the exhausted man. It may disturb a balance exquisitely adjusted and yet, in the main, counterpoise a scale heavily laden with disadvantages. If alcohol exhilarates, imparts comfort and energy, counteracts fatigue, hunger and unrest, then it does in effect, increase the capacity for work of those who take it under such circumstances ; and affords, in so far. a direct benefit and advantage."—*Dr. Brinton.*

—" Let it" (alcohol) "be taken never as a stimulant or preparation for work, but as a defence against the injury done by work, whether of mind or body. For example ; it is best taken with the evening meal, or after toil. Let the increase in the desire for and power of digesting food be the guide and limit to the consumption of all alcoholic liquids. Let the forms be such as contain the least proportion of fusil oil. Let all with an hereditary tendency to hysteria or other functional disease of the nervous system refrain from its use altogether, even though as yet they are in good health."—*Dr. T. King Chambers.*

—" All nations that have led the van in the march of civilization have been addicted to drink—ay, and addicted to drunkenness. The Jews, the Greeks, the Romans, the Germans, the Swedes, the Danes—not to mention the English, all round the globe, are amply attested by their own native literature to have been distinguished above their contemporaries in this way. It is true that some reactionary races, famous as conquerors, have been abstinent, but they and their faiths are dying out and the coloring they have given to civilization is even now fainter than that left by the robuster races a thousand years before they were heard of."—*Dr. T. King Chambers.*

— Referring to medium quantities of alcohol : " There is nothing, therefore, to prevent us from assuming that alcohol is a food, though as to its proper action we are yet in doubt ; so completely is it assimilated that no well defined traces of its products have so far been noted."—*R. A. Witthaus, M.D., Prof. of Physiology, Universities of New York and Vermont, in his " General Medical Chemistry," 1881.*

—" When diluted alcohol may be a food, a medicine, or a poison according to the dose. Taken in moderation with food it aids digestion."—*Prof. Witthaus, 1881.*

—" In this holy microcosm " (the human body) " are represented the kingdoms of nature, only immeasurably more perfectly. * * The next is the still more remarkable discovery of alcohol, representing wine. I hope, therefore, I shall be forgiven by those gentlemen who practise and preach total abstinence for saying that there is no such person as a teetotaller, in the literal and scientific sense of the word, on the face of the earth."—*J. P. Lewis, M.D., University of Brussels, L.R.C.P. Edin. ; M.R.C.S. Eng.*

—"After noting how some of Dr. B. W. Richardson's medical theories had been exploded. Dr. Lewis states his belief that Dr. Richardson's theory about alcohol depressing the system has just as little truth. He sums up ' I prescribe total abstinence as an extreme remedy for a desperate disease, but those that are whole need not a physician—nor the remedy.' "—*Prof. Sidney Ringer.*

—" To speak," as Dr. B. W. Richardson does "of the danger of producing such slight excitement of the hearts' action, is an abuse of science ; if it were true every exertion, every pleasing emotion, nay, all activity of every sort, would tend to shorten our days, and they alone would be wise and happy who exist in torpor."—*Dr. Greenfield.*

—" The use of alcohol in every age and by every nation in the world demonstrates that alcohol satisfies a natural instinct ; that it literally refreshes the system exhausted by physical or mental, labour and that it not only quickens the appetite for food and aids in digestion, but that it opens the digestive organs by limiting the amount of solid food which would otherwise be required. But in accomplishing this salutary end, it does not act as a mere condiment ; it is also a food ; in this sense at least, that it offers itself in the blood as a substitute for tissues which would otherwise be destroyed. ' Alcohol,' says Moleschott, ' is the savings bank of the tissues. He who eats little and drinks alcohol in moderation retains as much in his blood and tissues as he who eats more and drinks no alcohol.' It is certain that of any amount of alcohol ingested, not in excess of what the system requires, only a minute fraction can be recovered from any of the secretions ; the remainder undergoes metamorphosis and is eliminated in a new form. The notion that alcohol is unassimilated, and is chiefly eliminated through the lungs rested upon this peculiar fact that all the alcoholic beverages contain ethereal elements which are so eliminated, and thus the breath is affected. * * Alcohol is not exhaled by the lungs or skin, or excreted by the urine, unless the body be supersaturated with it."—*Anstie and Brintz, in U. S. National Dispensary, 1884.*

—" The first pre-requisite for the acquisition of a sound knowledge of the dietetics of the sick is to have clear ideas on the origin and meaning of the dietetic customs of the healthy ; for it is obvious that the proper diet for the sick must be some purposive modification of the diet of the healthy."—*Prof. Sir W. Roberts, Cardiff, 1885.*

—" It is important to remark that the main dietetic customs of all countries grow up and are established for the benefit of the robust and healthy, of the sober and temperate and those of mean average constitution ; in other words for those who are bearing the burden of the day and fighting the battle of life. These form the great mass and bulk of the adult population upon whose bodily and mental efficiency national progress and ascendancy depend. A good many individuals and even entire families may not find these customs in certain particulars beneficial to their exceptional tendencies or weaknesses. They may even find in them a source of destruction to their health and life, but here as elsewhere, and indeed universally in nature's operations the individual is sacrificed to the welfare of the community."—*Sir W. Roberts, Cardiff, 1885.*

—" In taking dietetic customs as objects of study, it is obvious that widely disseminated customs, followed by many races and by vast masses of population have a deeper and broader significance than customs limited to a few races or to small communities."—*Sir W. Roberts, Cardiff, 1885.*

—Referring to the dietetic customs of Britain and Western Europe, Prof. Roberts says : " Alongside the main dietetic habits formed for the operative mass of the community, there are secondary habits formed for the use of infants and children, and for persons advanced in years. With regard to infants and children we observe that they are not allowed to partake of the accessory articles of food which form so conspicuous a part of the dietary or their elders. They are allowed neither the use of alcoholic beverages nor of tea or coffee—except gradually as they draw toward adult age—but are fed on simple nutrients—milk, cooked cereals and more or less meat. With advancing years the diet undergoes certain modifications ; the consumption of meat is, I think, somewhat lessened, and the consumption of soups, milk, and cooked cereals proportionally increases."—*Sir W. Roberts, Cardiff, 1885.*

—" With regard to alcohol this modification of diet seems to vary with the preceding practice of the individual. Persons who have been in the habit, during their prime, of taking a full allowance of stimulants gradually diminish the proportion as age creeps on and their nutritive processes decline in elasticity and power. Sometimes the indications of this natural tendency are neglected or resisted by the unwary : they imagine that the quantity of stimulants they tolerated with impunity during the vigour of manhood cannot hurt them in later life. This is a serious mistake, the commission of which tends to accelerate senile decay, and to provoke fatally tending organic changes in the large organs of the body and in the arterial system. On the other hand, persons who, during their youth and prime have only used alcohol occasionally or have abstained entirely from it, find advantage in their declining years in a more systematic use of alcoholic beverages."—*Sir W. Roberts, Cardiff, 1885.*

—" Not very long ago those unhappy folk who go wearily and sadly because, forsooth, they are waxing fat, were warned to leave off drinking largely and to minimize the quantity of liquid they consumed. Never before, perhaps, was there a more mischievous fad imposed on a too credulous public than this reduction of the amount of fluid taken. * * Take, for example, the uric acid ; this excrementitious product requires not less than eight thousand times its bulk of water at the temperature of the blood to hold it in solution, and if it be not dissolved it rapidly crystallizes with more or less disastrous consequences, as in gout, gravel and probably many other less well recognized troubles. * * Three and a half pints should be consumed by any person in the twenty-four hours, and where the body is bulky four or even five pints should be the average."— *Annotation in London " Lancet," Nov. 25, 1885.*

—Referring to the increase of drunkenness in France, especially northern France, reported to the French Senate, an annotation in the London *Lancet* says :—" To remedy the evil it is proposed to increase the tax on alcohol and to impose more stringent regulations with respect to wine shops. The phylloxera is undoubtedly in a great measure responsible for the increase of drunkenness in France. Given cheap and pure wine, there would probably be but little alcoholism. Drunkenness is almost unknown in wine-growing countries, and intemperance in France prevails principally in the northern departments, where the wine cannot grow. What is more particularly wanted is the stringent application of laws against the sale of bad alcoholics and adulterated wine, combined with facilities of transit, so that pure wine may be brought within reach of the poorer classes. When once the palate is accustomed to natural and wholesome wines, the intoxicants that do so much harm will be considered too coarse to be drunk with any sort of pleasure. A Spanish peasant from the vineyards of Andalusia, accustomed to the unsophisticated Manzanilla or the Montilla of his native country would find the schiedam drunk in Northern Europe inexpressibly nasty. It is difficult to prevent drinking by act of Parliament, but the Legislature may seek to check fraud and adulteration."—*Annotation in London " Lancet," June 25, 1887.*

—"While the subjects were creating a demand for nourishment by exercise and starvation diet" (very moderate diet) "there was no ascertainable elimination of alcohol by the kidneys. When on the other hand the demand for nourishment was reduced to a minimum by rest and feeding, the elimination of alcohol by the kidneys was easily demonstrated." (The test was the chromic acid reaction.) * * "Unless our results have been rendered fallacious by errors of practical manipulation, which I can hardly think probable our experiments certainly indicate in the clearest manner that alcohol, in small doses at all events, is a food."—*Dr. T. W. Thompson, M.R.C.S., April, 1885.*

—"Almost all persons in fairly normal health may partake of sound and ripe fruit in greater or less amount. Among exceptions may be noted the gouty and rheumatic diathesis."

—"The physiological opinion opposed to these arguments is, that while alcohol, like other similar substances, has, in large quantities, a narcotic, a devitalizing effect, it has in small quantities a stimulating effect, between which and narcotism there is a difference, not of degree, but of kind. The stimulating effect is precisely the same with that of highly nutritious and easily digested food ; as regards the vital functions, it differs from the effect of ordinary food only in the rapidity of production. It does not substitute an abnormal for the normal action of the bodily organs ; it restores their natural functions; and it is capable of rectifying either deficient or redundant functional action. The only positive difference of effect between ordinary food and alcoholic stimulation is that the latter does not to any great extent add to the bulk of the body. There is no recoil or reaction after it, except that, as in the case of ordinary food, the effect is exhausted after a time. There is nothing to support the belief in a reaction, except the depression involved in the gradual recovery from the narcotic effect of a large quantity of alcohol, but between the narcotic effect of a large and the stimulating effect of a small quantity there is, as already said, a difference of kind—their connection is merely accidental. And the experience of mankind—the fact that moderate drinking does not usually pass into excessive drinking—sufficiently shows that it is not found necessary to increase the quantity. Since stimulation restores the natural functions, it, of course, is capable of removing the consequences of functions being perverted. Thus, it is maintained, that among other things, it gives relief from pain and muscular spasms, reduces the circulation when too rapid, produces healthy sleep, and removes general debility as well as the fatigue of the special organs. * * If these views are correct it follows that alcohol taken cautiously and in small quantities—the quantiites varying with the circumstances and with the constitution of the individual, may be used not only with safety but with advantage."—*Chambers's Encyclopædia.*

—"It is not disputed that many persons live in health without them--that persons having an abundance of wholesome food, not overworked, living in well constructed houses, and in wholesome air, can usually dispense with them. But when some or all of these conditions are wanting, which in towns at any rate, happens in all but exceptional cases, it is alleged that a nearer approach to health is made when a moderate use is made of alcohol."—*Chambers's Encyclopædia.*

—"Dr. Richardson 'of course' advocated total abstinence. We pronounce in favor of extreme moderation."—*British Medical Journal.*

—"Lewis Cornaro, who was a corpulent, gouty man, put himself upon a spare dietary, viz., twelve ounces of food, mainly vegetables, and fourteen ounces of light wine daily; and on this lived in good health for over half a century."—*Dr. Milner Fothergill.*

—"Some generous wine will very commonly be found to agree well with persons with old standing heart or lung mischief and will help them at their meals. In attacks of dyspnœa some alcohol as a stimulant is of incalculable value. As cordials alcoholic compounds are excellent."- -*Dr. Milner Fothergill.*

—" The Englishman who will persist in the dietary and especially ' the roast beef of old England' in tropical climates sooner or later falls ill with some hepatic trouble,"—*Dr. Milner Fothergill.*

—" There can be no doubt that a large proportion of the diseases of the digestive apparatus which are so fatal among European residents in India and other tropical climates result from the habitual ingestion of a much larger quantity of food than * * the system requires."—*Dr. W. B. Carpenter.*

—" Speaking of a dietary which embraces meat three times daily, two glasses of wine and two pints of ale, Dr. Milner Fothergill says :—' The dietary is that of well-fed, prudent navvy who diets himself so as to earn the best wages he can."

- " With persons of weak digestion, it is a good rule to eat less and to be particular about the food being easily digestible, when tired. Perhaps to eat less and take a little extra wine is the best plan to adopt."—*Dr. Milner Fothergill.*

—" Referring to the diet of lymphatic persons, Dr. Coombe says : 'Aromatics and spices, however, are useful, as is also wine in moderation and conjoined with adequate exercise."

—" The stomach of one man is offend' d and irritated by wine and his digestion impeded, whilst the appetite of another is improved and his digestion facilitated. The former is better without alcohol and he comes into the category of fools if he takes it ; but the latter has no claim to the character of physician if he abstains at the bidding of a mistaken fanatic or mere theorist."—*Dr. J. R. Bennett, LL.D., F.R.S., President Royal College of Physicians.*

—" I believe that alcohol has a special advantage over other articles of diet in restoring exhausted nervous power and repairing the waste that has taken place. I believe alcohol to be among the gifts of God accorded to man for therapeutic as well as other beneficial purposes—to make glad his heart and strengthen his nerves."—*Dr. J. R. Bennett.*

—" There are no statistics to prove that abstinence from the moderate use of alcohol is attended with unusual length of life or improvement of health. —*Alfred B. Garod, M.D., F.R.C.P., F.R.S.*

—In a sketch of the life of the late Prof. E. L. Youmans, published in the *Popular Science Monthly,* March, 1887, Prof. Youmans' sister thus refers to " Alcohol and the Constitution of Man," a book published by Prof. Youmans in 1853, and " based upon the view that alcohol is in all cases a brain poison." " The temperance people urged him to make a book of it, which he accordingly did, but further examination of the subject made him uncertain of his ground and the work was allowed to go out of print." This long since repudiated book is still, however, quoted in favor of universal total abstinence.

—" Grapes bear on their surface all that is necessary to cause saccharine water to ferment even when protected from the air.—*Schützenberger.*

—In Weston's walk of 100 miles in 22 hours 39 minutes—his best record —he drank, during the last part of the walk, three ounces of champagne and two and a half ounces of brandy. In his famous walk of 317½ miles in five days, his pulse rose from an average below 80 previous to the walk to 109 on the third day, fluctuating a great deal, and the temperature of his blood actually fell from 99°.5, the day previous to the walk, to 95°.3 on the first day and 94°.8 on the second.

—" There is good ground for believing that alcohol lessens the excretion of nitrogenous material from the body. It would seem most probable that

this is achieved by checking tissue waste, but it may be that alcohol * *
simply renders the digestion of the food more perfect, and lessens the production of food urea."—*U. S. Dispensatory, 1883.*

—"Alcoholismus chronicus is very rare in our hospital, because our labouring people drink beer. Distilled liquors are used very seldom. We
have had in the last year (1883) in our hospital nearly ten thousand patients,
but only twenty-one cases of alcoholism."—*Dr. Von Liemssen, Director of
the General Hospital of Munich.*

—"At one time in my life I regarded alcohol as the cause of half the cases
of insanity, because I had been taught that such was the fact.   Now I believe, of course speaking from my experience alone, it produces a very small
amount of such disease.   My error was and is a common one, and is one
into which both the profession and the laity fall."—*Dr. A. I. Thomas, Indiana Hospital for the Insane,* 1884.

—"However willing to admit drunkenness as a frequent source of physical disease, I very much doubt the truth of the reiterated assertion that it is
often the immediate cause of insanity, and still more of general paralysis."—
Dr. Thos. T. Austin, Medical Officer of the Bethnal House Asylum.

—"The number of insane persons admitted to this department of the hospital from the time it was opened in October, 1859, to the first of January,
1884, was 2,212. Of these 2,212 persons the insanity was caused by intemperance in 285 cases. Of these 285 the following number would sooner or
later have become insane from other causes, 56."—*S. Preston Jones, Med.
Supt. Pennsylvania Hospital for the Insane.*

—"This hospital was opened Nov. 1, 1848. From that date to Oct. 31,
1870, there were 4,431 persons admitted. In that number were 147 cases
attributed to alcoholism—143 men and 4 women."—*Wm. B. Fletcher,
Supt. Indiana Hospital for the Insane.* Supt. Fletcher's tables from 1871 to
1873 show 6,934 admissions, of which 195, or 2·8 per cent., are attributed
to alcoholism. In 1883 there were 698 admissions, of which 17, or 2·15 per
cent., are ascribed to alcoholism.

—In returns from 54 leading asylums of the United States, covering 36,-
973 patients, 2,588 cases are ascribed to alcohol, or 6·99 per cent. The number
thus classed embraces in the returns of many asylums cases where several
causes, including intemperance, co-operate.—1883.

—The admissions to the Bakkehus and Mathison Asylum for Idiots, 1871
to 1880, were 196, of whom 19 were born of intemperate parents.—*Danish
Bureau of Statistics, Report on Inebriety,* 1882.

—Of 500 cases treated in King's County, N. Y., Asylum for Inebriates
(pop. of county in 1880, 599,495), the necessity for medical treatment is attributable to distilled liquors in 441 cases, to distilled and fermented liquors,
in 35 cases ; to fermented liquors, in 24 cases.

—Warden Murray, of King's County, N. Y., Poor House, gives the history
of 671 indigent male persons supported at the expense of the county.   The
cause of dependence was : Physical disability in 457 cases ; want of work
in 99 cases ; intemperance in 72 cases ; vagrancy in 33 cases ; senility in 10
cases.   Nationality of persons whose intemperance caused pauperism :
Irish, 38 ; American, 28 ; German, 3 ; English, 2 ; Scotch, 1.

—"Is it not, indeed, probable that were all brain stimulants other than
ordinary foods common to man and other animals at once and forever annihilated or the alcoholic varieties alone withheld from common use, that the
result would be in the course of time deleterious to mankind by reason of
brain deterioration resulting from a loss of such food and a consequent gradual (no matter how slow) return of the races to a more common level at the
expense of those who have accomplished the greatest departures therefrom ?"
Dr. Everts.

de (Dr Wm. Sharpe) finds that the vine and the product of the vine have e en in olden times more intimately associated with man's intellectual growth and development than with his purely physical wants. The stimulus of alcohol when judiciously controlled ' always leads to active and higher mental efforts on the part of individuals,' thus producing a contrary effect to that of other stimulants, which tend rather ' to bring out a contented state of dreamy inaction,' and to repress effort. ' To understand fully,' he says, ' the beneficial action of alcohol as regards mental development we must first get a clear view of the value of those states of cerebral excitement which most people, though in varying degrees, experience something of, rising as they do, mentally above the level of what may be called their ordinary every day thoughts. * * The stimulus produced by alcoholic liquors, if not nearly of so high an order, is more easily called into play, while in a practical sense the latent ability being present it is more vigorous and effective as regards actual work. Hence the value of alcohol as a stimulant lies in the fact that it produces artificially and sustains temporarily that state of mental excitement or exaltation necessary to the conception and projection, though not to the detailed elaboration of those enduring works that whether in the domains of art, architecture or engineering, are remarkable for boldness of execution, originality and grandeur of design , and farther, that it is the only manageable stimulant which when used in moderation and in the form of wine or spirits is not only not injurious, but conduces to the general health while it favours both mental and physical development. Dr. Sharpe also assigns to alcohol a beneficial agency in stimulating genial thoughts and feelings."—*Popular Science Monthly, April*, 1884.

—The following remarkable records of pedestrianism are given in the latest Sporting manual. Match for championship of the world, $2,500 each, April 2-7, 1877 : D. O'Leary (non-abstainer), 519 miles 1585 yards—time, 141 h. 6 min. 10 secs.; E. P. Weston, 510 miles. P. Fitzgerald, (beer) 610 miles in 141 hrs.: Weston, 550 miles 110 yards in 141 h. 55 min. 10 secs.; Mark C. Rowell (non-abstainer) and E. P. Weston—Weston gave up at 198 miles, and Rowell stopped after covering 280 miles in 2 days 15 hrs. 17 min. Weston, 1977½ miles, resting Sundays, and walking on country roads, in 1000 consecutive hours. Within the past few years the notable feat of walking 5306 miles by Fitzgerald on beer against 5000 miles in 100 hours by Weston on water was accomplished.

—Dr. Koch, the noted German discoverer of the comma baccilus of cholera, is reported to state that the bacilla will not live in beer. During the last severe epidemic of cholera in Vienna it is stated that not one brewers' employee died.

—Dr. Anstie ("Stimulants and Narcotics") gives a detailed account of a man treated in Westminster Hospital who for about twenty years lived on nothing but gin and water (unsweetened) and a finger length of bread per day, generally toasted, and a few pipes of tobacco. His average daily allowance of gin was one bottle. He was eighty-three years of age when admitted to the hospital for treatment for a respiratory complaint.

—" Instances are not very uncommon of men drinking, for years in succession, as much as a pint of spirits and from half a gallon to a gallon of beer a day. The large majority of such persons, as far as my experience goes, live upon a very inadequate quantity of solid food, even when their exertions entail constant and strenuous bodily exertion "—*Dr. F. E. Anstie (Stimulants and Narcotics)*.

—" I have dwelt on this fact with regard to alcohol, and have detailed experiments which exemplify it, because I am anxious to impress on the reader's mind an important inference which appears to follow inevitably from it, namely, that alcohol was never designed by the wisdom of Providence to be employed by the human race *as an anæsthetic at all*, but for the sake of those stimulant qualities of its non-narcotic doses which are to a certain extent also shared by small doses of æther and chloroform. It seems as if the former were intended to be the medicine of those ailments which are engendered of the *necessary* every-day evils of civilized life, and has, therefore, been made attractive to the senses and easily retained in the tissues, and in various ways approving itself to our judgment as *a food* ; while the others which are more rarely needed for their stimulant properties, and are chiefly valuable for their beneficent temporary poisonous action, by the help of which painful operations are sustained with impunity, are in great measure deprived of these attractions and of their facilities for entering and remaining in the system."—*Dr. Anstie (Stimulants and Narcotics).*

—" I have never been able, for instance, to perceive the logical necessity of the inference that because a large excretion of carbonic acid in the breath coincides with the digestion of certain undoubted foods, the alimentary value of the latter was to be considered as directly proportionate to such increase."—*Dr. Anstie (Stimulants and Narcotics).*

—" The intemperance of the teetotallers has greatly hindered the cause of temperance. Their denunciations of moderation and moderate drinking have estranged whole classes of men who would have brought strength and help to their cause. Their dogmatism about what constitutes food and what poison and about the injurious effects of every drop of alcohol has excited ridicule and laughter rather than intelligent interest. By such an attitude the medical profession has been prevented from co-operation with those who see in the drunkenness of the country a gigantic evil and disgrace, to be remedied at any cost short of that of talking nonsense."—*Editorial in " London Lancet,"* 1884.

—" I think we have a number of cases of which mania è *potu*, or acute mania from drink is the type, in which the alcohol acts as an excitant of morbid cerebral function. In these cases there is almost invariably a strong hereditary tendency, or a previous history of brain disease ; and the alcohol lights a train of mischief already laid, and which might have been exploded by any other moral or physical cause of mental excitement."—*Dr. Bucknill,* 1878.

—" Consider the great part which grief and anxiety, worry and overstrain, play in the production of insanity, the depressing effects of poverty and the failing struggle for existence, of misery in all its forms, and then consider to how great an extent the use of alcohol oftentimes tends to make the burthen of life bearable, if not by stimulating the powers, at least by deadening the sensibilities of men ; and I think you will agree with me that by the occasional help of strong drink a man may sometimes be able to weather that point of wretchedness upon which his sanity would otherwise have been wrecked."—*Dr. Bucknill (Habitual Drunkenness and Insane Drunkards.)*

—" Alcohol in its physiological action, is *atriptic*, retarding the disintegration of the tissues, especially of the nerve tissue ; and when the brain is wearing itself into madness, alcohol, at the right time, and in the right doses, without doubt, sometimes checks the ebb-tide of reason."—*Dr. Bucknill (Habitual Drunkenness and Insane Drunkards.)*

form
in th
throt
scho
ciou:
ada

BENJAMIN WARD RICHARDSON, M.A., M.D., LL.D., F.R.S., F.R,C,P

Honorary Physician to the Royal Literary Fund and Author of the Cantor Lectures on Alcohol.

The acknowledged head of the medical teetotallers is Dr. B. W. Richardson, whose Cantor lectures, delivered in 1874-5, have, through republication and industrious circulation by various teetotal organizations, been widely disseminated throughout Anglo-Saxondom, and have been utilized on platform and in pulpit and press in the service of prohibition. A text-book for use in the schools, by the same author, shortly followed the Cantor lectures, and through the efforts of various teetotal organizations has been forced into the schools of several American States and Canadian Provinces under the specious pretence of being "Scientific Temperance." The text-book is an adaptation of the Cantor lectures. Both works are of an extraordinary character : the methods adopted suggest more than the mere error and one-sidedness into which an extreme man might fall. In order to support universal total abstinence the effects of excess are by inference transferred to temperate use, as if the action of different quantities was the same in kind and differed only in degree—a most fallacious idea. Certain physiological actions are exaggerated and misrepresented, while incidental counteractive actions are ignored. Physiological phenomena pertaining to the action of many kinds of food, in common with alcohol, and generally beneficial in their moderate manifestation, are represented in excessive exhibition with a view to suggesting false inferences. In fact, the physiology of these books is very similar in kind to that of a certain well-known class of alarmist quack advertisements, which fulfil their purpose of impressing the ill-informed. Were the same kind of reasoning applied to common articles of diet many people would eat their meals in fear and trembling, for there is scarcely anything which may not be, with some constitutions, and with most people under certain circumstances, even when indulged in in moderation, productive of serious injury. Salt, beef, vinegar, could by the methods cleverly and ingeniously, if unscrupulously, applied in these books, be shown to be most injurious articles of diet, capable of leading to dire results. Sugar could, and with a measure of reason, be credited, even where used in comparative moderation, with producing fatty degeneration of the heart, liver and kidneys, apoplexy, paralysis, indigestion, nervous irritability, insanity, crime and misery, and the medical evidence to make a case strong enough to impress and even convince millions that it is a dangerous article of diet would be readily forthcoming were it called for to aid in the propagation of a popular moral "fad." Rest, exercise, study, heat, cold, cou'' all be represented in an alarming aspect, simply by the application of methods analogous to those adopted by Dr. Richardson, who has been brought into deserved ridicule in consequence in the scientific world, and whose notoriety, achieved almost entirely by teetotal laudation, is mistaken by many for fame.

The errors and perversions—historical, physiological, and in regard to fact —the false coloring and false suggestion, the suppression of important truths, and the direct misstatements to be found in Dr. Richardson's books, would require longer space for discussion than our limits will permit. The direct contradiction of many of these errors will be found in reading the great physiologists of the age, while many will be at once apparent to intelligent men even without much acquaintance with physiological science. We append a few extracts from Dr. Richardson's "Temperance" Text-Book as examples of the misstatement, silliness and prejudice sought to be imposed on our children in the name of "Scientific Temperance," leaving it to the common sense of the reader to judge whether an author capable of penning these extracts can have the disposition or capacity to carefully weigh facts and theories in such a manner as to make his opinions of any scientific value.

[Note—The words in brackets are not in the Text-Book.—Ed]

—" In all my experience I never once knew a person who liked the first taste of any one of the drinks we are now thinking about. This fact seems to me to show clearly that it was never intended that human beings should take these drinks regularly every day."—p. 8. (Oysters, tomatoes—bah ! A limited experience only could fail to notice that many persons, even children, relish wine at first taste, and some other liquors as well.)

—" Although there are so many drinks made and sold as beers, wines and spirits none of them are fitted to the first natural wants and desires of man. I gather from the facts before us that the said drinks are therefore not wanted at all."—p. 8. (Nor for the same reason tea, coffee, &c.?)

—" If a little child can live, and grow up, and learn, and work, and play, and be very healthy, and pretty, and strong, and happy without these drinks a man and woman can live without them equally well."—p. 8. (Applied to other things, how would this logic answer?)

—" It is right to remember that the same rule applies to all the lower animals, and that, too, through every period of their lives. * * Many other animals that are not mammals have never at any time any drink except water."—p. 9.

—" To think of these facts is to feel the best of proof that artificial fluids have no place whatever in the scheme of creation, and that the natural human instinct, which, as we have seen, abhors at first artificial drinks, does so because it would if it had its way lead men through the same simple process of living as it does the less endowed animals."—p. 10. (That is to say, we should "go to grass.")

—" Besides the colloidal and starchy and fatty foods which the water helps to bring into solution, there are other substances which we take in, by or with foods, and which are called salts. Common table salt, chloride of sodium, is a good example of this nature. These salts are very soluble in water, and the water is the means of conveying them in solution into the blood."—p. 15. "Alcohol acts on the blood in somewhat the same manner as salt does, and gives a tendency to a disease similar to that which is produced by living a long time on salted provisions, to which disease the name of scurvy is commonly applied."—p. 51.

—" Living bodies are built for water only to be the fluid by which they may live. * * Even those animals which are nearest to man in their appearance and anatomy (monkeys and apes) even these have always continued to exist on water. It is man alone who by the aid of his unfettered reason and skill has changed the natural ordinance."—p. 18. (And man alone who uses a carriage instead of walking as the animals do.)

—" Moreover, amongst those people who have invented different drinks there have been none who have not mainly subsisted on water. No one has yet lived, actually lived, on any other fluid."—p. 19.

—" The worst persons who break the natural law retain water as their staple drink, though many are so blindly ignorant that they do not know the

fact. They drink what they think is not water at all, and they give to the drink so used many fine names. They pay for the fluids they drink very heavy prices. They go to a great deal of trouble in order to obtain them. But strangest of all is that in spite of fine names, expense and trouble, they only succeed at last in getting a drink the larger part of which is water. *
* Poor people ! why at the very time they are holding up a drink which has in it at least three parts out of four of water."—p. 19.

—" From the earth the water rises in distillation under the heat of the sun. Ascending in vapor, it becomes pure in the air.   It condenses on the mountain tops from the vapourous into the liquid form, and streams down in rivulets and torrents.   It condenses in clouds and falls in showers of rain. It is stored up for us in cool reservoirs of the earth or runs to us in rivers or brooks."—p. 21.   (This is correct, and agrees with all the geographies.)

—" I hope these *important truths* are now understood.   First, that wine, which is water containing a new liquid, did not come down to man flowing as pure water flows from the earth, a liquid ready made as drink for all animals, including man."—p. 23.

—" Secondly, that the liquid which, added to water, makes wine does not exist ready made as water exists, springing from the earth."—p. 23.

—" Thirdly, that in order to obtain the fluid which so changes water as to transform it into wine it is necessary to let the juices of fruit *mix with water* and undergo fermentation."—p. 23.   (This about mixing grape juice with water is not true, but teetotallers are not supposed to know anything about vineyards.)

—" And that if this process were ot *constantly* being performed in the *most wholesale* way by men who *devote their lives to the work* there *would be no supply of wine or strong drink at all.*"—p. 23.   (It is difficult to properly ᵓᵗ aracterize this statement without using words which might suggest uncom- ᵓᵗnentary reflections on either the mental or moral condition of the ardent ᵓᵗles of Dr. Richardson, who are responsible for bringing such ᵓᵗ import- ... ᵓᵗths " before the minds of our pupils, and so imbuing them with high respect for " Temperance Science " and advancing " the cause of truth and righteousness."   Of course, children might not know how natural and simple is the making of wine and beer, or the full meaning of the words " home-brewed ale " and " home-made wine," nor might they be aware of where the rural populations of wine and beer countries now derive, and still more, in all past ages before the subdivision of labour resulted in the common establishment of wine vaults and breweries, did derive their supplies of beer and wine.   And so on the assurance of a text-book they might grow up in bliss-ful ignorance into good prohibitionists, who would imagine that abolishing breweries, &c., would easily secure the abolition of alcoholic drinks, and the cause of morality would thus be " promoted.")

—" When the ancients drank wine at their feasts they mixed it with water, and at first used it very sparingly." (When it was scarce.—ED.) " It was offered at the sacrifices to the gods, and *young men under thirty were not permitted to take it at all.   Women were not permitted to take it at any age.* In time these restrictions were broken through."—p. 24. (It is almost incon-ceivable that Dr. Richardson did not know that the prohibitions spoken of existed only in a few localities at most, such as Rome, and only for a limited period, and that therefore the ascription of these prohibitions to the " an-cients " generally, as is implied in the text, is unwarranted.)

—" A pigeon will take, without showing the slightest symptoms, as much opium as would kill several men.   A goat will swallow, without injury, a quantity of tobacco which would kill several men.   A rabbit will swallow, without injury, a dose of belladonna that would kill several men. But neither the pigeon, nor the goat, dor the rabbit can swallow alcohol without being

influenced by it much in the same manner as a man would be."—p. 41.
(Ergo—terrible poison, isn't it?)

—" As soon as the taking of alcohol becomes a habit either in a lower animal or in a man the quantity required to produce obvious effects has, as a
rule, to be increased."—p. 42. (What is meant by " obvious "? Science has
conclusively proven that when taken in moderate or *stimulating* quantities
it is not necessary to increase the quantity to secure the same effects. It is
only when immoderate or *narcotic* quantities are frequently used that a
physiological necessity for an increase exists.)

—" She (Nature) did not give man boiling water to drink and leave him to
discover that the water was dangerous unless it were cooled to a certain safe
degree. (Sic ') Neither did she give him such active chemical fluids as ardent spirits for his nourishment. Yet there are many persons who declare
that alcohol is a food."—p. 44. (Therefore hot oast beef, frozen water, hot
water, &c., &c., &c., should not—what ?)

—" We can drink milk without being burned."—p. 44.

—" Compared with milk alcohol shows no trace of being a food in any particular. We want no better evidence, and we could have no better proof that
alcohol is not a food."—p. 44. (Sic.)

—" The warmth of surface which seems to be imparted by alcohol only
seems to be imparted. Positively the warmth is not imparted by alcohol, but
is set free by it. * * Alcohol acts in the same manner as cold."— pp. 60,
61. ◆(This argument, intended to ridicule people who say in ordinary well
understood language that alcohol warms them, is a silly " catch," and if in
a sense correct it might also be applied as follows : The warmth which seems
to be imparted to a boy's hand by a strap only seems to be imparted. Positively the painful warmth is not imparted by the strap but is set free by it.
The man who gets his skin into a radiating glow by brisk exercise and says
he is warmer is foolish. The warmth only seems to be imparted. He is
really bringing the warmth to the surface of his body, and by exposing it to
radiation is setting free the warmth.—This may in some circumstances, just
as is the case with alcohol, be objectionable, but in others is distinctly promotive both of comfort and physical benefit. The cup of warm tea on a cold
day may also in like manner be beneficial or not according to the circumstances under which it is taken. That alcohol does produce warmth, or its
equivalent, is a necessary deduction from scientific law.)

—" The alcohol, from its great affinity for water, induces those who drink
it to imbibe water to excess. * * Alcohol when it is freely diluted acts as
I have said on the fibrine of the blood, rendering the fibrine unduly fluid. In
fact, alcohol acts on the blood in somewhat the same manner as salt does."
—p. 51. (So water, like alcohol, can actually be taken to excess, and result
injuriously. The statement in the first sentence is apt to provoke a smile
amongst the vast majority of users of diluted alcohol, who find no such result. The whole explanation of scorbutic conditions associated with these
extracts is grossly misleading.)

—" A disease of a similar kind (to scurvy) sometimes affects poor people
who live on land, and who are obliged for the sake of economy to subsist on
coarse bread, bacon, or other salted foods, and who try to make up the deficiency by resorting to beer or perhaps to stronger drinks. If they would
* * take to plain and nourishing foods, like oatmeal and peasemeal, they
would be very much better off in mind, body and estate."—p. 52.

—" All great consumers of alcohol are chillier during the winter months
than are those who abstain."—p. 71. (And very frequently those who abstain are chillier than those who use alcohol in such moderation as tends to
increase bodily health.)

—(The argument of the chapters on heart action, stimulation and depresrion, nervous action, respiration, &c., is characterized by the plausible use of some actual facts, exaggeration and distortion of others, and careful suppression or ignoring of still other facts, and by false suggestion, while throughout there is a steadily marked endeavour to confound the results of excess, often explained in a misleading way, with the effects of moderation. And finally, there is a ghastly array of diseases presented as the results of alcohol, diseases in most cases quite as readily provoked by the wrong use of other things rightful in themselves. And there is a total omission of numerous morbid conditions remedied, modified or prevented by the use of alcohol.)

—" Insanity—It is certain that no single cause of madness is so frequent." —p. 94. (Dr. Workman's reports of Toronto Asylum show only a very small percentage ascribed to drink, and twice as large a percentage ascribed to religious excitement. Recent returns from American asylums show about 7 per cent. or less than from each of several other causes—or occasions, more strictly speaking,—of insanity.)

—" We shall, however, still find many to defend the use of alcohol, for many, very many, are still ignorant about it ; many, very many, are strongly prejudiced in favour of it ; many, very many, are so fond of it that they cannot help praising it as a good thing for themselves, and therefore as a good thing for everybody. Such is the strange perversity of the human mind, that numbers of people who are going wrong, and who know they are going wrong, in the use of alcohol will still persist in their error, and with their eyes open to the wrong they are doing will persist in leading others with them. It is one part of the madness inflicted by alcohol on its friends that it deceives them and in turn makes them deceivers."—p. 95.

—" The reason why those who are habituated to the use of alcohol believe that moderate quantities of it inflict no harm is that they do not clearly know what it is to be free from alcohol."—p. 77.

—" You will have often in your lives to listen to the arguments of these persons. They will tell you a great deal of error, which you must be ready to hear and at once recognize as error."—p. 95.

—" If men could have been kept sober by practising temperance the fight against intemperance would never have had to be fought."—p. 78. (And if men could have been kept honest, unselfish, &c., by practising temperance in all the things of life, the fight against avarice, cruelty, excess of passion, gluttony, &c., and unrighteousness generally, would not have to be fought, and we would not have societies actually existing, or which might be suggested, proposing to abolish private property, ribbons, kid gloves, pies, and other things which many people make the occasion of sin. Nor would Communism, proposing the entire abolition of the right of private property, have as many adherents in Europe as the parallel absurdity—prohibition—has in America.)

———

—(" It will be asked : was there no evidence of any useful service rendered by the agent in the midst of so much obvious evidence of bad service ? I answer to that question that there was no such evidence, and there is none. * * It is an agent as potent for evil as it is helpless for good."—*Dr. B. W. Richardson—Action of Alcohol on the Body.*)

National Liberal Temperance Union.

✱

Literature Committee.

✳